GROUND STUDIES FOR PILOTS
Volume 3

By the same authors:

Ground Studies for Pilots, third edition
Volume 1, Radio Aids
Volume 2, Plotting and Flight Planning

Aviation Law for Pilots, third edition

By H. A. Parmar:

Navigation General and Instruments
(published by H. & J. Parmar)

GROUND STUDIES FOR PILOTS

Third Edition
Volume 3

NAVIGATION GENERAL

S. E. T. Taylor

formerly British Airways and Chief Ground
Instructor, Malaysia Air Training, Kuala
Lumpur; Chief Ground Instructor, London
School of Flying

H. A. Parmar

Senior Tutor, Bristow Helicopters Ltd
formerly Chief Ground Instructor, Malaysia
Air Training, Kuala Lumpur; Specialist
Instructor, London School of Flying

GRANADA
London Toronto Sydney New York

Granada Publishing Limited — Technical Books Division
Frogmore, St Albans, Herts AL2 2NF
and
3 Upper James Street, London W1R 4BP
866 United Nations Plaza, New York, NY 10017, USA
117 York Street, Sydney, NSW 2000, Australia
100 Skyway Avenue, Rexdale, Ontario, M9W 3A6, Canada
PO Box 84165, Greenside, 2034 Johannesburg, South Africa
CML Centre, Queen & Wyndham, Auckland 1, New Zealand

ISBN 0 246 11177 1

First published in Great Britain by Crosby Lockwood & Son Ltd 1970
Reprinted 1972
Second edition 1974, reprinted 1976, 1978
Third edition, in three volumes, 1979

British Library Cataloguing in Publication Data
Taylor, Sydney Ernest Thomas
Ground studies for pilots; — 3rd ed.
Vol 3: Navigation general
1. Airplanes — Piloting
I. Title   II. Parmar, Hasmukhlal Amritlal
629.132'5216   TL710

ISBN 0—246—11177—1

Text set in 10/12 pt IBM Press Roman, printed and bound
in Great Britain at The Pitman Press, Bath

Granada ®
Granada Publishing ®

Contents

Preface

This third and concluding volume of *Ground Studies for Pilots* includes all the subjects which are commonly referred to as navigation general — maps and charts, aircraft instruments, magnetism and compasses, and the solar system. They have been completely revised and brought into line with the latest syllabusi of both professional pilots' licences, and we hope have been even further clarified without making them less readable.

We are grateful to HMSO for permission to reproduce two pages of the Air Almanac.

January 1979

S. E. T. Taylor
H. A. Parmar

Section 1

MAPS AND CHARTS

1: Mathematics Reminders

Mathematics plays an important part in the professional pilot's ground training syllabus: a solid background of elementary mathematics with an ability to shift figures quickly not only helps a student with his exams, but is a necessity in understanding a great deal of the syllabus material. You have done it all before, have forgotten much of it, or are just feeling shaky about it: in this short chapter, we remind you of some of the pertinent maths processes without offering any text book proofs or heavy explanations.

Assuming that without a certain facility with fractions, and decimals (since without that you couldn't have got this far), we will start with:

Ratio and proportions

A ratio is the mathematical relationship between two quantities. It indicates the number of times one quantity is contained in another quantity of the same kind, and is either expressed as a fraction, x/y or simply $x:y$.

Example. An aircraft's maximum speed is 180 kt whereas its cruising speed is 150 kt. Then the ratio of the cruising speed to the maximum speed is $\frac{150}{180}$, that is, $\frac{5}{6}$. Working in reverse, if the ratio between any two quantities and the value of one of them is known, then the value of the other may be found. In the above example, when the ratio of 5/6 and the cruising speed are known, the maximum speed is expressed in the relationship

$$\frac{5}{6} = \frac{150}{\text{max. speed}}$$

Thus
$$\text{max. speed} = \frac{150 \times 6}{5} = 180.$$

Here is another application. Divide 84 in the ratio 3:4. Add 3 and 4 to give 7, divide 84 in seven parts, that is, each part is equal to 12. Then 3 parts is equal to 36 and 4 parts is equal to 48, the two shares adding up to the original figure of 84.

On many occasions we handle ratios without being conscious of it. The Greek letter π that we use in maths is the ratio of the circumference of a circle to its diameter. When we talk of a one million chart we mean a ratio of 1:1 000 000. An aircraft's mach number is the ratio of its true airspeed to the local speed of sound.

When two ratios are equal, then the four quantities comprising these ratios are said to be in proportion and the whole term may be expressed in the form $a:b::c:d$. The numbers 3, 4, 18 and 24 form a proportion. The ratio $\frac{3}{4}$ is equal to the ratio $\frac{18}{24}$. If we know any three of these terms the fourth may be found very simply. For example in an expression similar to that above, the first term is 4, the third 12 and the fourth 36. Then the second term, b, stands in relationship to the rest as

$$\frac{4}{b} = \frac{12}{36}$$

Then $$12b = 4 \times 36$$

$$b = \frac{4 \times 36}{12} = 12.$$

If one quantity B depends on another quantity A so that when the value of A increases, the value of B similarly increases, then B is said to be directly proportional to A and is expressed as $B \propto A$. Conversely, if the value of B decreases as the value of A is increased, B is said to be inversely proportional to A and is written as $A \propto 1/B$. In the above example, the terms $4:12::12:36$ are in direct proportion. If 4 kg of fuel takes you one ground nautical mile and 8 kg takes you two ground nautical miles then the terms are in direct proportion. Alternatively, if one workman takes one full day on a job whereas two would do it in half a day, the two quantities are in inverse proportion. Problems involving direct and inverse proportions are solved as follows.

Example.   If an aircraft travels 25 nm in 4 min, what distance will it cover in 16 min?
The distance covered in 1 minute is the ratio $\frac{25}{4}$.
∴ in 16 minutes the distance travelled $= \frac{25}{4} \times 16$ nm
$$= 100 \text{ nm}$$

Example.   A survival dinghy carries sufficient emergency rations to last four survivors five days. How many days will the rations last if there were six survivors aboard?
No. of survivors is increased in the ratio $\frac{6}{4}$.
∴ no. of days is decreased in the ratio $\frac{4}{6}$.
Hence the total number of days $= \frac{4}{6} \times 5 = \frac{20}{6} = 3\frac{1}{3}$.

Some examples

1.   What ground distance in nm is represented by 7.296 in on a half-million map?
1 in on the map represents 500 000 in on the ground.

∴ 7.296 in represents $\dfrac{500\,000 \times 7.296}{6\,080 \times 12}$ nm

$$= 50 \text{ nm.}$$

2.   In the time that an aircraft A travels 240 nm, another aircraft B travels 270 nm. How far has B travelled when an aircraft A has gone 320 nm?
When A travels 240 nm, B travels 270 nm.
∴ ratio B to A is $\frac{270}{240}$ or $\frac{9}{8}$.
When A has gone 320 nm,
B has travelled $\frac{9}{8} \times 320 = 360$ nm.

3.   If a $2°$ change of latitude gives sunrise 5 min 12 s earlier, how much earlier will sunrise take place for a change of latitude of $3°30'$?
$2°$ change lat. gives sunrise 5 min 12 s earlier.
∴ $1°$ change lat. gives sunrise 2 min 36 s earlier.
∴ $3°30'$ change lat. gives sunrise 2 min 36 s x 3.5 = 9 min 06 s.

Percentages

A percentage is simply a fraction with a denominator of a 100, the words per cent meaning 'per hundred'. The fraction $\frac{69}{100}$ expresses 69 as a percentage and may be rewritten 69%. Any other fraction may be expressed as a percentage by multiplying it by 100. Thus, $\frac{6}{10}$ when expressed as a percentage is

$$\frac{6}{10} \times 100 = 60\%.$$

A decimal fraction may be converted to a percentage by moving the decimal point *two* places to the right. For example, $0.039 = 3.9\%$ and $5.4 = 540\%$.

Example. A pilot flying at 6 000 ft increases his altitude to 7 400 ft. What is the percentage increase of the altitude?

The ratio of the increase is $\dfrac{1\,400}{6\,000}$

and the percentage increase $= \dfrac{1\,400}{6\,000} \times 100 = 23.3\%.$

To express quantity a as a percentage of another quantity b, multiply the fraction a/b by 100.

Example. Express 120 kg of fuel as a percentage of 760 kg of fuel.

$$\frac{120}{760} \times 100 = 15.78\%.$$

Practice problems

1. Due to head winds an aircraft is required to increase its normal route fuel from 10 000 kg to 11 845 kg. What is the percentage increase in fuel?

$$\frac{1\,845}{10\,000} \times 100 = 18.45\%.$$

2. If the route fuel is 23 670 kg and 7% of this is added as a contingency measure, how much fuel is carried?

Start fuel 23 670 kg

$$7\% \text{ of start fuel} = \frac{7 \times 23\,670}{100}\,\text{kg}$$
$$= 1\,656.9\,\text{kg}$$

∴ Fuel + contingency = (23 670 + 1 657) kg = 25 327 kg.

3. When flying at a given mach number, if the TAS at FL 380 is 400 kt and at FL 320 is 415 kt, what percentage increase in the TAS is achieved by flying at the lower flight level?

$$\frac{15}{400} \times 100 = 3.75\%.$$

4. On a Mercator chart a true ground distance of 274 nm measures 271 nm. What is the percentage error?

$$\frac{3}{274} \times 100 = 1.09\%.$$

Square roots

Until the use of electronic calculators is allowed in the examinations, this old-fashioned method of finding square roots must be used. And this is one mathematical process which is quickly forgotten on leaving school.

Let us find the square root of the number 136 161. The first step is to divide the number into groups of two, starting from the right. The groups are 13 61 61. Notice that the first group happens to have two digits in it (13); if the number has an odd number of digits you will have only one digit in the first group.

By trial, find the number whose square will exactly equal the first group and in the event of there not being one, then the one which comes closest to it. 3 x 3 is 9 and 4 x 4 is 16. So the number we are looking for is 3. 3 is the first figure in your answer. Then take away 9 from 13. Adopt the following layout with the answer at the top

$$
\begin{array}{r}
3 \\ \hline
/13\ 61\ 61 \\
-9 \\ \hline
6/\ \ 461
\end{array}
$$

The remainder is 4 and bring down the next group. The number now is 461. Multiplying 3 (first figure in the answer) by 2 gives 6 which is the first digit in the term which will divide 461.

By trial find a number to go after 6 so that when the two numbers are multiplied by that number the product is 461 or nearest to it. The best approach is to see how many times 6 goes into 46 (ignore 1) and try that. It goes 7 times and 67 x 7 = 469. The product slightly exceeds 461 and therefore the figure we are looking for is 6. 66 x 6 = 396. Enter 6 after 3 in the answer and take away 396 from 461 to repeat the process. The sum at this stage looks like this:

$$
\begin{array}{r}
36 \\ \hline
/13\ 61\ 61 \\
-9 \\ \hline
66\ \ \ 461 \\
-396 \\ \hline
72\ /\ 6561
\end{array}
$$

Repeating the process, 36 x 2 = 72 gives us the first two figures in the divisor of 6 561. By trial, 729 x 9 = 6 561 and the final answer is 369. The complete sum should look like this:

$$
\begin{array}{rl}
369 & \qquad \text{Answer} \\ \hline
/13\ 61\ 61 & \\
-9 & \\ \hline
66\ /461 & \\
-396 & \\ \hline
729\ /6561 & \\
-6561 & \\ \hline
/0000 &
\end{array}
$$

In finding the square root of a decimal number, the grouping is slightly different. Start at the decimal point, and split into groups of two from left and right, e.g. 316.895 is grouped as 3 16 . 89 5. To complete the pair in the decimal part, a 0 is added to the 5, making the grouping . 89 50. The process of finding the square root is exactly as already outlined. In the solution of this particular question, you could carry on *ad infinitum*, adding a couple of noughts and getting a vast number of decimal points in the answer.

Radians

The radian is the angle at the centre of a circle subtended by an arc of length equal to the radius of that circle. Although not used extensively in aviation it is useful to be familiar with this unit. As the circumference of a circle is expressed as $2\pi R$ (where R is the radius), in a complete circle we have 2π radians. If an angle is known in degrees, the same angle in radians can be found from the relationship

$$\text{radians} = \text{degrees} \times \frac{\pi}{180}.$$

Conversely, to convert from radians to degrees use the relationship

$$\text{degrees} = \text{radians} \times \frac{180}{\pi}.$$

And lastly, the length of arc an angle subtends is given in the formula

radius of circle x angle in radians.

Algebra

We will now revise the type of problem that crops up, which we know some students have lost their youthful ability to solve readily.

Simple equations are a nothing: the popular sort in this business go:

How far can an aircraft go from base at a G/S out of 450 kt and back at 400 kt, if fuel flow out is 6 600 kg/hr and back is 6 100 kg/hr, with usable fuel on board of 40 000 kg?

Distance divided by G/S out gives *time* on the outward leg.
This time multiplied by fuel flow gives *fuel used out*.
Ditto for the homebound leg.
The total of the two fuels equals the FOB.

Algebraically, where x is the distance in nm, this gives

$$40\,000 = \left(\frac{x}{450} \times 6\,600\right) + \left(\frac{x}{400} \times 6\,100\right)$$
$$= \frac{44x}{3} + \frac{61x}{4}$$
$$480\,000 = 176x + 183x$$
$$\therefore \quad x = 1\,337 \text{ nm}.$$

Not overpoweringly difficult, but simultaneous equations often give difficulty.

Simultaneous equations

To solve simultaneous equations

(a) multiply or divide the coefficient of either unknown (as convenient), ignoring the sign, so as to make it equal in both the equations.

(b) Then add or subtract the two equations. This process eliminates that unknown quantity whose coefficients were made equal.

(c) With only one unknown quantity left in the equation, solve for that unknown quantity.

(d) Finally, substitute the value thus found in either of the original simultaneous equations and solve for the other unknown.

Example $2x + 3y = 13$ ...(1)
                             $x - 2y = -4$ ...(2)

Multiply the first equation by 1, and the second by 2

$$2x + 3y = 13$$
$$2x - 4y = -8$$

Subtract $+ 7y = +21$
                             $y = 3$

Substitute for y in equation 1

$$2x + 9 = 13$$
∴ $x = 2$

Remember it now!

Indices

A quick reminder before we go on to logarithms.
To multiply — add the indices: $a^m \times a^n = a^{m+n}$.
To divide — subtract the indices: $a^m \div a^n = a^{m-n}$.
To raise to a power — multiply the indices: $(a^m)^n = a^{mn}$.
To find a root — divide the indices: $\sqrt[n]{(a^m)} = a^{m/n}$
And oddments:

$$\sqrt{a} = a^{\frac{1}{2}}$$
$$\sqrt[3]{a} = a^{\frac{1}{3}}$$
$$\frac{1}{\sqrt{a}} = a^{-\frac{1}{2}}$$
$$a^2 b^{-2} = \frac{a^2}{b^2}$$

Logarithms

Once you have found the log of a number, you must decide its *characteristic*: this is the whole number part, for what you have extracted from the tables is the decimal part, the *mantissa*. If the number you have logged is less than 10, then there is no characteristic: thence, for each place beyond the first, the characteristic is increased by 1, e.g. 21 would have a characteristic of 1, 210 a characteristic of 2, 2 100 a characteristic of 3. The characteristic is one less than the number of digits before the decimal point, in other words.

Numbers less than 1 which are to be logged have a negative characteristic: −1 for the decimal point, and a further −1 for each zero after the decimal point, e.g.

                0.535 would have −1 (expressed as $\bar{1}$)
                0.0 535 would have $\bar{2}$
                0.00 535 would have $\bar{3}$.

To multiply two numbers, find their logs and add them.
To divide two numbers, find their logs and subtract them.
Look up the result in the anti-log tables (or in the body of the log tables) and extract the pure number. Adjust it according to the characteristic. You do *not* include the characteristic when you enter the anti-log tables — only the mantissa. But

in adding a brace of logs, say, they are treated algebraically, including the characteristic, e.g.

multiply 52.86 x 0.329 x 6.835.

Number	Log
52.86	1.7231
0.329	$\bar{1}$.5172
6.835	0.8347
	2.0750

Look up .0750 in the anti-log tables. This gives 1189 which with a characteristic of 2
becomes 118.9.
One more example:
divide 0.001567 by 0.0005316.

Number	Log
0.001567	$\bar{3}$.1951
0.0005316	$\bar{4}$.7256
	0.4695

You will notice that the log subtraction is treated as simple algebra. Look up .4695
in the anti-log tables. This gives 2947, and the answer is 2.947.

Powers and roots. To raise a number to a given power, multiply the logarithm of
the number by the power. Similarly, to find the root of a number, divide the
logarithm of the number by the root.
A pair of examples won't come amiss:

$$2^4 = 4 \log 2$$
$$= 4 \times 0.30103$$
$$= 1.20412$$

antilog = 1600, which with a characteristic of 1 is 16,
and find the cube root of 125.

$$\log 125 = 2.09691$$

divide by 3, 2.09691 ÷ 3

$$= 0.69897$$

Antilog is 5.

The following is occasionally set in ATPL Met.

The MSL pressure is 1016 millibars (mb), temperature 12°C, and the
temperature at the 700 mb level is 2°C, what is the true height of the 700 mb level?

The formula is

$$h = 221.1T (\log p_1 - \log p_2)$$

where T is the mean temperature absolute in the air column, p_1 is the pressure at
MSL and p_2 is the pressure corresponding to height h.

Temperature 12°C to 2°C = 7°C mean
$$= (273 + 7) \text{ absolute}$$
$$= 280 \text{ absolute.}$$

$p_1 = 1016$ mb, and $p_2 = 700$ mb.

$\therefore h = 221.1 \times 280 (\log 1016 - \log 700)$

Now watch it:

∴
$$\log 1016 = 3.00689$$
$$\log 700 \ \ = 2.84510$$
$$\overline{\hspace{2cm}0.16179}$$

You must now *re-log* 0.16179, as it were, to solve the equation
$$\log 221.1 \ \ = 2.34459$$
$$\log 280 \ \ \ \ \ = 2.44716$$
$$\log 0.16179 = \bar{1}.20895$$
$$\text{Add} \hspace{2cm} \overline{4.00070}$$

The anti-log is 10016.
The true height of the 700 mb level is 10016 feet.

Properties of a circle
A few facts:

(a) Circumference of a circle = $2\pi r$ (where r is the radius).

(b) Length of an arc = $\dfrac{N}{360} \times 2\pi r$ (where N is the number of degrees in the angle subtended by the arc at the centre of the circle).

(c) A perpendicular bisector of a chord from the centre of a circle divides the chord in two equal parts.

(d) Equal chords are equidistant from the centre of a circle.

(e) Angles in the same segment of a circle are equal.

(f) The angle at the centre of a circle equals twice the angle at the circumference (fig. 1.1).

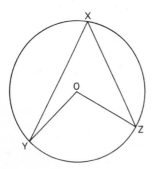

Fig. 1.1

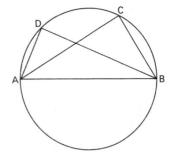

Fig. 1.2

Angle YOZ is twice the angle YXZ.

(g) The angle in a semi-circle is a right angle.

In fig. 1.2, the straight line AB divides the circle in two equal parts. Segment ADCB is therefore a semi-circle and the angle ACB is a right angle. So is the angle ADB.

Trigonometry
Consider the following right-angled triangle:

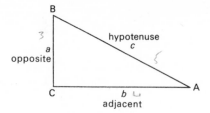

Fig. 1.3

Check

$$\text{Sin A} \quad = \frac{\text{opp}}{\text{hyp}} = \frac{a}{c}$$

$$\text{Reciprocal Cosec A} = \frac{\text{hyp}}{\text{opp}} = \frac{c}{a}$$

$$\text{Cos A} \quad = \frac{\text{adj}}{\text{hyp}} = \frac{b}{c}$$

$$\text{Reciprocal Sec A} \quad = \frac{\text{hyp}}{\text{adj}} = \frac{c}{b}$$

$$\text{Tan A} \quad = \frac{\text{opp}}{\text{adj}} = \frac{a}{b}$$

$$\text{Reciprocal Cot A} \quad = \frac{\text{adj}}{\text{opp}} = \frac{b}{a}$$

These ratios are constant for a given angle. When an angle exceeds 90°, the required ratio is first given a sign according to fig. 1.4. Here the ratios and their reciprocals are

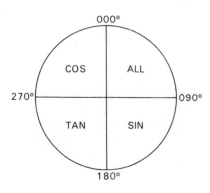

Fig. 1.4

positive in the given quadrants. Elsewhere they are negative. Then the angle is reduced to less than 90° before the tables are entered, viz:

When the angle θ is in the *second* quadrant, it becomes (180° − θ).

When the angle θ is in the third quadrant, it becomes (θ − 180°).

When the angle θ is in the *fourth* quadrant, it becomes (360° − θ).

Example.   Find the sine of 210°.

210° is in the third quadrant, ∴ equivalent acute angle = 210° − 180° = 30°. Sine of

$30°$ is 0.5 and the sign is $-ve$; therefore the value is -0.5.

In a triangle ABC, $\angle C = 90°$, $a = 32.3$, $b = 25.9$. Find $\angle A$.

$$\text{Tan } A = \frac{a}{b} = \frac{32.3}{25.9}$$

No.	Log
32.3	1.5092
25.9	1.4133
	0.0959

0.0959 is the logarithm of the tangent of angle A. From the log tan tables, we extract angle $51°17'$ for the numerical value of 0.0959. Thus, angle A is $51°17'$.

For solving triangles other than right-angled triangles there are two standard formulae:

(a) $$\frac{a}{\sin A} = \frac{b}{\sin B} = \frac{c}{\sin C}.$$

(b) $a^2 = b^2 + c^2 - 2bc \cos A.$

This formula may be rewritten as

$$b^2 = a^2 + c^2 - 2ac \cos B$$
$$c^2 = a^2 + b^2 - 2ab \cos C.$$

1.   In a triangle ABC, $a = 421.9$, $c = 321.6$, $\angle A = 65°12'$, find $\angle C$.

Answer: $43°47'$.

Solution

Two sides and one of the opposite angles are given, therefore we will use the simpler formula (a) above.

$$\frac{a}{\sin A} = \frac{c}{\sin C} \quad \text{or} \quad \frac{\sin C}{c} = \frac{\sin A}{a}$$

Thus, $\sin C = \dfrac{c \sin A}{a}$

$$= \frac{321.6 \sin 65°12'}{421.9}$$

$$= 43°47' \text{ (from log sine tables).}$$

No.	Log
321.6	2.5073
$\sin 65°12'$	$\bar{1}.9580$
	2.4653
421.9	2.6252
	$\bar{1}.8401$

2.   In a triangle ABC, $a = 12.8$, $b = 15$, $c = 9.9$, find $\angle C$.

Solution

Using the formula from (b) above

$$c^2 = a^2 + b^2 - 2ab \cos C$$
$$-2ab \cos C = c^2 - a^2 - b^2$$
$$2ab \cos C = a^2 + b^2 - c^2$$
$$\cos C = \frac{a^2 + b^2 - c^2}{2ab}$$
$$\cos C = \frac{12.8^2 + 15^2 - 9.9^2}{2 \times 12.8 \times 15}$$

	No.	*Log*	*Antilog*
$a =$	12.8	1.1072	
		×2	
$a^2 =$		2.2144 ⟶ 163.9	+
$b =$	15	1.1761	
		×2	
$b^2 =$		2.3522 ⟶ 225.0	
$a^2 + b^2 =$			388.9

$c =$	9.9	0.9956	
		×2	
$c^2 =$		1.9912 ⟶ 98.0	−
$a^2 + b^2 - c^2 =$ numerator =			290.9

2		0.3010
$a =$	12.8	1.1072
$b =$	15	1.1761

$2ab$ = denominator = 2.5843

numerator 290.9		2.4637
denominator		2.5843−
cos C		1.8794 ⟶ 40°45′

Answer: $\angle C = 40°45'$.

2: Form of the Earth

The earth

In maps and charts the cartographers attempt to reproduce, as faithfully as possible, a part or a portion of the earth's surface. Therefore, before we study the individual projections, it is essential that we first take a look at the form of the earth itself and see how a position fixing graticule of latitude and longitude is created.

The earth is not a true sphere. It is flattened slightly at the poles and its shape is variously described as an 'ellipsoid', 'oblate spheroid' or in desperation simply a 'geoid' which means 'earthlike'. This flattening of the poles is known as 'compression' and it is the ratio of the difference between the equatorial diameter and the polar diameter to the equatorial diameter. It is expressed as

$$\text{compression} = \frac{\text{equatorial diameter} - \text{polar diameter}}{\text{equatorial diameter}}.$$

If we put the appropriate figures in the formula we can soon find the value of this ratio. The equatorial diameter being 6 883.7 nm and the polar diameter, 6 860.5 nm, the ratio becomes:

$$\frac{6\,883.7 - 6\,860.5}{6\,883.7} = \frac{23.2}{6\,883.7} = \frac{1}{297} \text{ or } \frac{1}{300} \text{ near enough.}$$

What all this means in practice is that if we built a model of the earth on a reduced scale (called the 'reduced earth') so that its equatorial diameter was 300 inches, then the model's polar diameter would be 299 inches.

This shows that in producing maps and charts for normal usage a refinement for the earth's shape is neither necessary nor profitable. As for navigation charts, no errors of any significance occur through lack of refinement, and therefore, the remainder of the discussion on maps and charts is based on the assumption that the earth is a perfect sphere.

On any plain sphere, in order to establish a reference system it is necessary to adopt a convention defining the directions and a location-fixing system of co-ordinates. These are established as follows:

The poles

The poles are situated on the earth's surface at the ends of the rotational axis. The north pole is that end of the earth's rotational axis about which the earth rotates anti-clockwise when viewed from above it. The south pole is the opposite end of the rotational axis to the north pole. The two poles locate the extreme north and south points.

The equator

The equator is a great circle (discussed later) with its plane perpendicular to the earth's axis. Since the earth's axis defines the north—south direction, a plane perpendicular to it must define the east—west direction. Thus, the plane of the equator lies in the east—west direction. All points on the equator are equidistant from the poles and it divides the earth into the northern and southern hemispheres.

Latitude

A latitude is defined as the *arc of the meridian* (see fig. 2.1) intercepted between the place and the equator. It is measured in degrees, minutes and seconds north or south of the equator. One degree (1°) equals 60 minutes (60′) and one minute equals

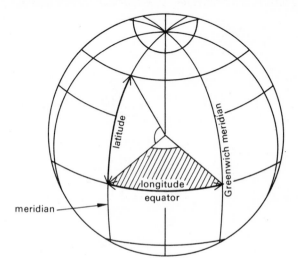

Fig. 2.1

60 seconds (60″). The choice of the equator as a datum of measurement is a natural one, on account of its having the largest circumference of all the parallels of latitude.

The planes of all other latitudes lie parallel to the plane of the equator. The arc of the circle from the equator to either pole being a quarter of a circle, the maximum possible latitude is 90° north or south, see fig. 2.2. The spacing of the parallels of latitudes along the meridians is perfectly uniform. This means that the arc distance between, say, 10°N and 20°N is exactly equal to the arc distance between 70° and 80° or any other two parallels having a difference of 10°. This being so, the distance between any two parallels up a meridian may be calculated quite simply. A meridian together with its anti-meridian completes one full circle, that is, 360°. The circumference of any circle is given in the formula, circumference = $2\pi r$, where r is the radius. For the earth, the arc distance between any two parallels may be given in the relationship

$$\frac{\text{difference in latitude}}{360} \times 2\pi r.$$

From the above, and taking the radius of the earth to be 3 441 nm, the distance between two latitudes one degree apart is

$$\frac{1}{360} \times 2 \times 3.14 \times 3\,441 \text{ nm}$$

= 60 nm   along a meridian.

Finally, all parallels of latitude indicate an east–west direction. Thus, if you are flying along a parallel of latitude, you are flying a 090° or 270° track.

Longitude

A longitude is defined as the shorter *arc of the equator* (see fig. 2.1) intercepted

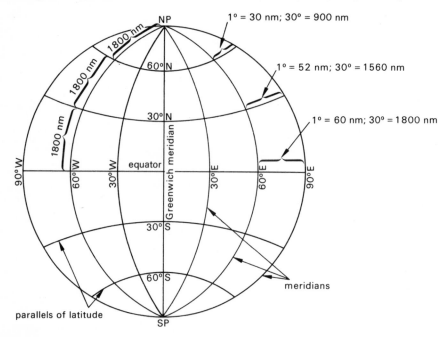

Fig. 2.2

between the Greenwich meridian and the meridian of the place. It is measured in degrees, minutes and seconds east or west of the Greenwich meridian. The anti-meridian of Greenwich is the maximum longitude possible, i.e. 180° E/W.

The term 'meridian' may be described as a semi great circle which runs north–south from pole to pole. Meridians, together with their anti-meridians complete the great circle. Every point on the earth has its own meridian passing through it. All meridians indicate north–south direction, the true north reference datum being along any meridian towards the north pole.

It will be noticed that with regard to the meridians we have no natural datum to start the count with and consequently, the meridian of Greenwich is arbitrarily taken as the datum. It is also called the 'prime meridian'.

The meridians are also placed uniformly along the parallels, the distance between any two meridians *along the equator* being

$$\frac{\text{difference in longitude}}{360} \times 2\pi r$$

For a one degree difference in longitude along the equator, the distance between the meridians is

$$\frac{1}{360} \times 2 \times 3.14 \times 3\,441 \text{ nm} = 60 \text{ nm}$$

which, you will notice, is identical to the distances between parallels along a meridian.

However, unlike the distance between two latitudes along a meridian, the distance between two given meridians along a latitude away from the equator does not remain constant at 60 nm for a 1° change of longitude. This is because, as can be seen in fig. 2.2, the meridians converge towards the poles. They are a maximum distance apart at the equator and converge into a point at the poles.

But taking another glance at fig. 2.2 it will be noticed that the lengths of the parallels of latitude also reduce (it reduces in proportion) once away from the equator. The circle of the equator, having the largest circumference gives the longest distance, whereas the latitude of 90°N/S is a mere point.

Therefore, the distance between the meridians along any parallel depends on the length of the parallel itself. Now, any quantity which varies from being maximum (that is, one unit) at 0° to a value of 0 at 90° in fact varies as a cosine curve. Accordingly, the variation in the lengths of the latitude follows a cosine relationship. Let us examine this.

The length of the equator, that is, the circumference of the earth is 360 × 60 nm = 21 600 nm. At 30°N/S the circumference is (21 600 × cos 30°) nm = 18 706 nm. Along this parallel, 360 meridians at one degree intervals are placed uniformly at a distance of $\frac{18\,706}{360}$ nm = 52 nm. Putting this another way, one degree at the equator contains 60 nm, and one degree at 30°N/S contains (60 cos 30°) nm = 52 nm.

Similarly, the circumference or the length at 60°N/S is (21 600 × cos 60) nm = 10 800 nm, and the distance between each meridian one degree apart is $\frac{10\,800}{360}$ nm = 30 nm. Or, (60 cos 60°) nm = 30 nm and this one fact, that the distances on the earth at 60°N/S are exactly one half of those at the equator, is worth remembering.

To summarise so far:

(a) All parallels of latitude are parallel to the plane of the equator and indicate the east—west direction.

(b) The spacing of the parallels along the meridian is given by

$$\frac{\text{difference of latitude (d lat)}}{360} \times 2\pi r.$$

(c) One degree difference of latitude along a meridian measures 60 nm anywhere between the equator and the poles.

(d) One degree difference of longitude (d long) measures 60 nm at the equator. Away from the equator this distance reduces at the rate of the cosine of the latitude and is found from the formula

> distance between two meridians one degree apart = 60 cos lat.

(e) All meridians indicate north–south direction.

Definitions

A few more definitions, terminologies and the facts now.

Great circles

If you draw the largest possible circle on a tennis ball, the circle's circumference will be the same as the circumference of the ball, its radius will be the same as the radius of the ball and in fact, it will divide the ball in two equal parts. What you have just drawn is a great circle.

A great circle is defined as a circle on the surface of a sphere whose centre and radius are those of the sphere itself, thus the plane of the great circle passes through the centre of the sphere, dividing it in two equal parts. Note these fundamentals arising from the concept of a great circle.

(a) Since it has the greatest radius possible, it has the least curvature and nearly approaches being a straight line.

(b) And, since a straight line between any two points is the shortest distance between those two points, we can say that the shorter arc of a great circle passing through any two points on the earth's surface represents the shortest distance between those two places.

(c) Only one great circle can be drawn through any two given points *unless* those two points are diametrically opposite in which case an infinite number of great circles can be drawn. An example is the meridians. A meridian, together with its anti-meridian is a great circle and any number of meridians can be drawn between the two poles (which are diametrically opposite).

(d) The equator is a great circle.

(e) The meridians (semi great circles) cut the equator at a constant angle of 90°. Elsewhere, a great circle cuts the meridians at a changing angle. This can be visualised by drawing a straight line in fig. 2.2 at any place other than along the equator. To fly the shortest distance between the two places we must fly the great circle track, but the changing angles makes such a flight quite difficult, unless flying due north/south along a meridian (fig. 2.3).

(f) A radio wave detected by an aircraft's receiver will have followed the shortest distance across the earth's surface. We must remember this when plotting such a radio signal on a chart on which a straight line is not a great circle.

Small circles

A small circle is any circle on the surface of the earth whose centre and radius are not those of the earth itself. Any parallel of latitude, other than the equator, is a small circle.

Rhumb line

A rhumb line (R/L) is a regularly curved line on the surface of the earth which cuts all meridians at the same angle, fig. 2.3. From this definition it will become apparent

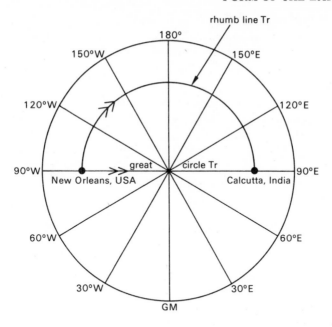

Fig. 2.3

that an aircraft flying a constant track will be following a rhumb line track. The other points to note are these.

(a) Only one rhumb line can be drawn through any two points.

(b) A rhumb line does not represent the shortest distance between two places, but it is convenient to follow.

(c) The meridians and the equator are the only examples of great circles which are also rhumb lines.

(d) All latitudes and longitudes intersect each other at 90°. Therefore, all parallels of latitude are rhumb lines. When flying along a latitude, you are flying a rhumb line track, make no mistake about it.

Change of latitude (ch lat)

This is the same as 'difference of latitude (d lat)' and the terms are interchangeable. It is defined as the arc of meridian intercepted between the parallels of the two places and is named north or south according to the direction of the change. Thus, a flight from 5°N to 20°N represents a ch lat (or d lat) of 15°N whereas a flight in the reverse direction changes its latitude by 15°S.

Change of longitude (ch long)

This, again, is the same as 'difference of longitude (d long)' and is defined as the smaller arc of the equator intercepted between the meridians of the places. It is named east or west according to the direction of the change. A change from 178°W to 178°E is a change of 4°W and the converse is true. If you can't visualise the other side of the globe, look in an atlas and make yourself familiar.

Distances

Now that we have established a convention for directions and a position locating
graticule we will turn our attention to the various units of distance measurement.

Nautical mile

A nautical mile is defined as the distance on the surface of the earth which subtends
an angle of one minute of arc at the centre of the earth, see fig. 2.4. Any distance on

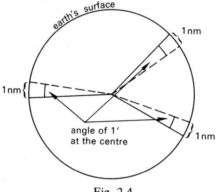

Fig. 2.4

the surface of the earth formed by an angle originating at the earth's centre is the
great circle distance. Now, it will be remembered that a latitude is also similarly
defined, that is, one minute of latitude is the arc distance on the surface subtended
by an angle of one minute at the earth's centre. This fact makes it possible to calibrate
the meridians in nautical miles i.e. 1′ of latitude is one nautical mile and 1° is 60 nm
when measured along the meridians.

However, because of the ellipticity of the earth the actual distance of one nautical
mile on the surface of the earth varies. In fact, because of this, the earth's centre is
variously defined as 'geographical centre', 'geocentric' and so forth, according to the
purpose of the definition. To illustrate the effect of the ellipticity, the shape of the
earth in fig. 2.5 is grossly exaggerated. In this figure the earth's axis is extended

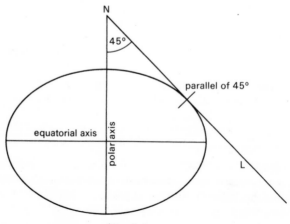

Fig. 2.5

northward to point N so that a straight line, NL, drawn from N makes an angle of 45°
to the axis. It can be shown by geometry that the point where this straight line is
tangential to the surface of the earth defines the parallel of 45°. Because of the shape
of the earth, that is where the parallel of 45° lies; if the earth was a perfect sphere,
45° would lie exactly half way between the equator and the pole.

The distance from the equator to 45° is 2 700 nm (45 x 60 nm) and there is the
same difference between 45° and the pole. Consequently, as can be seen in the figure,
a nautical mile nearer the equator is smaller than that at the pole.

One nautical mile at the equator measures 6 045.7 ft, at 45°N/S, 6 078 ft, at the
poles, 6 108 ft. For navigational purposes the standard length of a nautical mile is
taken as being 6 080 ft. The international nautical mile as defined by ICAO measures
1 852 metres.

Kilometre
A kilometre is 1 000 metres. A metre is 1/10 000 000th part of the average distance
along the earth's surface from the equator to either pole. A kilometre therefore equals
1/10 000th part of the average distance along the earth's surface between the equator
and the pole. For practical purposes it is taken as 3 280 feet.

As the distance between the equator and either pole is 5 400 nm and also 10 000 km,
it follows that 54 nm = 100 km.

Statute mile
Established by the Royal Decree of Queen Elizabeth I, the standard length of the
statute mile (sm) is 5 280 feet.

The conversions from one unit to another are as follows.

$$66 \text{ nm} = 76 \text{ sm}$$
$$41 \text{ nm} = 76 \text{ km}$$
$$41 \text{ sm} = 66 \text{ km}$$
$$1 \text{ nm} = 1.1 \text{ 515 sm}$$
$$1 \text{ nm} = 1.85 \text{ km}$$

Questions
If you have assimilated the contents of this chapter you should be able to answer
these questions.
1.   A pilot wishes to fly from A(50°N 16°E) to B(51°N 15°E). He has a choice of
the following two routes:
    (a) fly along the parallel 50°N until the meridian 15°E is intercepted and then fly
up the 15°E meridian to his destination; or
    (b) fly along the meridian 16°E until the parallel of 51°N is intercepted and then
fly along the 51°N parallel to his destination.
    Is there any advantage between the two routes? Give a reason for your answer.
2.   Give an example of a rhumb line which is also a great circle.
3.   Define (a) a great circle; (b) a rhumb line.
4.   Which parallel of latitude is also a great circle?
5.   Which small circles are also rhumb lines?

6.   What is the distance from Calcutta (30°N 90°E) to New Orleans, U.S.A. (30°N 90°W) over the north pole?

Answers
1.   Route (b) is shorter in the ground distance on account of the earth's convergency.
2.   Any meridian together with its anti-meridian is a great circle and since it cuts itself at zero degrees (a constant angle) it is also a rhumb line (see p. 19).
3.   See the text.
4.   The equator.
5.   Any parallel of latitude.
6.   7 200 nm.

3: Convergency, Conversion Angle, Departure

In this chapter we continue our study of familiarisation with the earth.

Convergency
We have already seen that a great circle cuts successive meridians at different angles because the meridians converge towards each other to the poles. This inclination

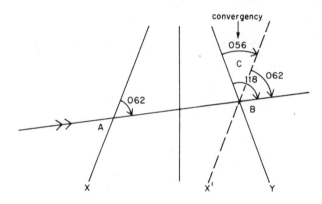

Fig. 3.1

between any two meridians is called *convergency*, and it equals the angular difference between the measurements of the great circle at each meridian.

The inclination of the meridians or convergency between two meridians X and Y (in fig. 3.1 above) may be found by various methods, as follows:
1.   By transferring on the chart the meridian X to X′ (or the other way round) to pass through position B, and measuring the angle formed by the two meridians Y and X′ at B (marked convergency angle in the figure).
2.   By measuring the angle that the straight line joining A and B makes at positions A and B and calculating the difference. In above figure, $\angle C = 118 - 062 = 056°$.
3.   By mathematical calculation, using formula

$$\text{Convergency} = \text{ch long} \times \text{sine lat.}$$

The formula is derived from the reasoning that the meridians cut the Equator at $90°$ and therefore, they are parallel to each other and the angle of inclination between them is 0. At each pole every meridian on the face of the Earth meets in a point. The convergency or inclination of any two meridians here therefore amounts to the change of longitude between them. For example, two meridians $1°$ apart will make an angle of 1° at the Poles. Therefore, at any intermediate latitude, the convergency must equal change of longitude x sine of the latitude.

This formula is correct only when two places A and B are on the same latitude. If A and B are at different latitudes, the sine of mid-latitude between them may be used in the formula, giving sufficiently accurate results for our purpose.

Great Circle – Rhumb Line Relationship

A straight line drawn on the Earth or on a chart having correct convergency (that is, chart meridians inclining at the same angle as the meridians at the same place on the Earth) will be a great circle. The associated rhumb line will appear as a curve. (An important point is illustrated here also: the rhumb line, or to be precise the tangent to it, at the central meridian is parallel to the great circle; when using charts with converging meridians in practice, the Rhumb Line Track is the same as the mean Great Circle Track, measured at the central meridian between leaving and arriving meridians.)

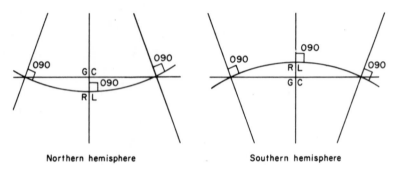

Northern hemisphere   Southern hemisphere

Fig. 3.2

However, on a chart which projects meridians as straight lines parallel to each other, a straight line will be a rhumb line and not a great circle (since it will cut all the meridians at the same angle). The example is the Mercator's projection, and on such charts the great circles must appear as curves (fig. 3.3).

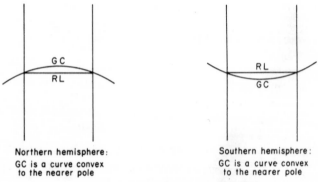

Northern hemisphere: Southern hemisphere:
GC is a curve convex   GC is a curve convex
to the nearer pole   to the nearer pole

RL in each case cuts the meridians at 90°

Fig. 3.3

If you observe figs. 3.2 and 3.3 you will notice that in either hemisphere the rhumb line, as regards the great circle, always appears towards the equator, a point useful to remember when plotting radio position lines.

The angular difference between the great circle and the rhumb line is called conversion angle (see *Ground Studies for Pilots* Volume 2, Plotting and Flight Planning) and its value is half the value of the convergency.

To prove that Conversion Angle (CA) = $\frac{1}{2}$C (convergency)

AB and CD are two meridians (fig. 3.4) and the straight line EF is a great circle.

$$\angle FEG \text{ is Conversion Angle and} = \angle EFH$$
$$= \angle JFK$$

Transfer AB to A_1B_1 to pass through position F. Then $\angle CFA_1$ is Convergency angle.

A rhumb line makes the same angle at each meridian,

$$\therefore \angle AEF + \angle FEG = \angle CFA_1 + \angle A_1FK - \angle JFK$$

But $\angle AEF = \angle A_1FK$

and transferring $\angle JFK$ to the left hand side of the equation,

$$\angle FEG + \angle JFK = \angle CFA_1$$

or $CA + CA = C$

that is, $2CA = C$ or Conversion angle = half the convergency.

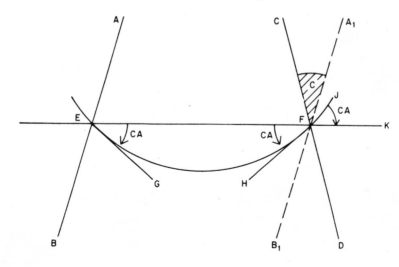

Fig. 3.4

Great Circle and Rhumb Line Bearings

Problems of the following nature are solved by application of the above theory.

1.   A bears 055° RL from B. What is the RL bearing of B from A? Since here we are only dealing with rhumb line bearings, B's bearing from A must be reciprocal of A's bearing from B. Therefore the answer is 235°.

2.   A bears 090° GC from B. Convergency = 4°. What is the GC bearing of B from A in the Northern and Southern hemispheres?

It is always advisable to draw up a simple sketch as shown below:

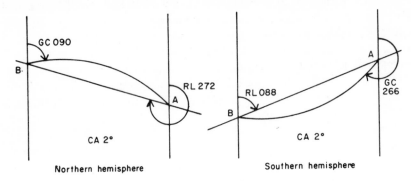

Fig. 3.5

Convergency given is 4°; therefore CA = 2°
A's RL bearing from B = 092; and B's RL bearing from A = 272°
∴ B's GC bearing from A = 274°
Southern Hemisphere: CA = 2°; A's RL brg = 088°; B's RL brg. from A = 268°
∴ B's GC bearing from A = 266°

3.   GC bearing of B from A is 095°; RL bearing of A from B is 273°.
(a) What is the CA?
(b) In which hemisphere are we?
(c) What is the GC bearing of A from B?

Solution
(a) CA is the angular difference between GC and RL
∴ CA = 095 + 180 = 275 − 273
= 2°

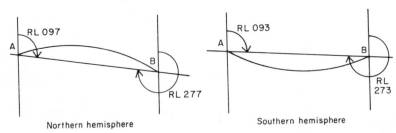

Fig. 3.6

(b) In fig. 3.6 above, for the Northern Hemisphere, the RL bearing of A from
B = 095 + 2 + 180
= 277°
and therefore the aircraft cannot be in the Northern hemisphere. The sketch
for the Southern hemisphere illustrates the facts as given.
(c) Since the RL bearing of A from B is 273, the GC bearing
= 273 − 2 (CA)
= 271°.

A further illustration:

Calculate the convergency between A (48°30′N 28°12′E) and B (55°00′N 10°00′E). If the Rhumb line Track between A and B is 300°, what is the initial Great Circle Track?

Solution

$$\text{Convergency} = \text{ch long} \times \sin \text{mean lat}$$
$$= 18°12' \times \sin 51°45'$$
$$= 18.2° \times .7853$$
$$= 14.3°$$

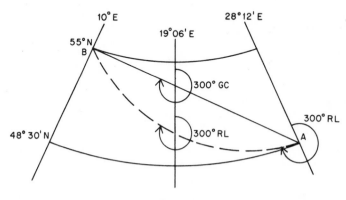

Fig. 3.7

The GC and RL Tracks are the same at the mid-meridian, that is, in this problem, at 19°06′E.

The total convergency between A and B is 14°, therefore the convergency between the mid-meridian and the meridian of A must be 7°; with the RL Track of 300°, the GC Track from A is 300° + 7° = 307°, and the simple sketch clarifies the solution.

In our experience we have found that some students initially find this topic a bit difficult to understand and apply. But you will soon get conversant with it: ample exercises are given here to help you. These exercises are graded so that they become progressively more involved from the simplest one until all principles are covered. When attempting the problems, know the different ways in which convergency/conversion angle may be found and keep the two formulae firmly in mind. Also remember that the convergency formula has three elements in it: convergency, longitude and latitude. Knowing any two of these the third one can be found. For example:

$$\text{difference of longitude} = \frac{\text{convergency}}{\sin \text{mean lat}}.$$

Exercise 1

1. Give ch long (change of longitude) between
    (a) 10°E and 20°E
    (b) 10°E and 160°W
    (c) 8°W and 8°E

(d) 160°W and 158°E

(e) 175°E and 168°W.

Answers: (a) 10°E; (b) 170°W; (c) 16°E; (d) 42°W; (e) 17°E.

2. If the RL bearing of A from B is 070°, what is the RL bearing of B from A?

Answer: 250°.

3. If the RL bearing of B from A is 300°, what is the RL bearing of A from B?

Answer: 120°.

4. GC bearing of Q from P is 060°, CA = 4°; if the two places are in the northern hemisphere what is the

(a) RL bearing of Q from P?

(b) RL bearing of P from Q?

(c) GC bearing of P from Q?

Answers: (a) 064°; (b) 244°; (c) 248°.

5. The GC bearing of X from Y is 315°; CA = 4°; if the two places are in the northern hemisphere what is the

(a) RL bearing of X from Y?

(b) RL bearing of Y from X?

(c) GC bearing of Y from X?

Answers: (a) 311°; (b) 131°; (c) 127°.

6. X and Y are in the southern hemisphere. X bears 120° RL from Y. If the CA is 2° what is the

(a) GC bearing of X from Y?

(b) RL bearing of Y from X?

(c) GC bearing of Y from X?

Answers: (a) 122°; (b) 300°; (c) 298°.

7. The GC bearing of M from N is 040°; the RL bearing of M from N is 042°;

(a) what is the conversion angle?

(b) in which hemisphere are the two places?

(c) what is the RL bearing of N from M?

(d) what is the GC bearing of N from M?

Answers: (a) 2°; (b) northern hemisphere; (c) 222°; (d) 224°.

8. The GC bearing of A from B is 340°; the RL bearing of A from B is 345°;

(a) what is the conversion angle?

(b) in which hemisphere are the two places?

(c) what is the RL bearing of B from A?

(d) what is the GC bearing of B from A?

Answers: (a) 5°; (b) southern hemisphere; (c) 165°; (d) 170°.

9. The GC bearing of B from A is 280°; the RL bearing of A from B is 096°;

(a) what is the conversion angle?

(b) in which hemisphere are the two places?

(c) what is the GC bearing of A from B?

Answers: (a) 4°; (b) northern hemisphere; (c) 092°.

10. The RL bearing of P from Q is 050°; the GC bearing of Q from P is 228°;

(a) what is the conversion angle?

(b) in which hemisphere are the two places?

(c) what is the GC bearing of P from Q?

Answers: (a) 2°; (b) southern hemisphere; (c) 052°.

Exercise 2

1. The GC bearing of A from B is 160°; if the CA = 2°, what is the
    (a) GC bearing of B from A in the northern hemisphere?
    (b) GC bearing of B from A in the southern hemisphere?
Answers: (a) 344°; (b) 336°.

2. The GC bearing of P from Q is 130°, the GC bearing of Q from P is 318°.
    (a) what is the conversion angle?
    (b) in which hemisphere are the two places?
    (c) what is the RL bearing of Q from P?
Answers: (a) 4°; (b) northern hemisphere; (c) 314°.

3. If the GC bearing of N from M is 280° and the GC bearing of M from N is 094°, what is the RL bearing of M from N?
Answer: 097°.

4. A and B are on the parallel of 30°N and A is at 8°W. If the GC bearing of B from A is 087° what is B's longitude?
Answer: 4°E. (Hint: all parallels of latitudes indicate E—W direction and all latitudes are rhumb lines.)

5. A's position is 41°26′S 175°W. The convergency between A and B is 7°. If B is due west of A and sin of 41°26′ is 0.7,
    (a) what is the GC bearing of B from A?
    (b) what is B's longitude?
Answers: (a) 266½°; (b) 175°E.

6. A's position is 54°S 160°E; the convergency between A and B is 21°. If B is due east of A and the sin of 54 is 0.8,
    (a) what is the GC bearing of B from A?
    (b) what is B's longitude?
Answers: (a) 100½°; (b) 174°W.

7. X is at 40°N 12°E; Y is at 40°N 06°E. If you are given that sin 40° = 0.64,
    (a) find the GC bearing of X from Y
    (b) find the GC bearing of Y from X.
Answers: (a) 088°; (b) 272°.

8. The earth convergency between two positions on the same parallel 8° of longitude apart, is 7.2°. On the same parallel, X is at 173°W and Y is at 176°E.
    (a) what is the GC bearing of X from Y in the northern hemisphere?
    (b) what is the GC bearing of Y from X in the southern hemisphere?
Answers: (a) 085°; (b) 265°.

9. A and B are both on the parallel of 30°. The longitude of A is 10°W. The GC bearing of A from B is 266°(T). What is the longitude of B?
Answer: 6°E.

10. A flight is made from VOR A (51°N 01°W), local variation 8°W, to VOR B (51°N 06°W), local variation 9°W. The radials are maintained throughout the flight. If the drift is 7° starboard give (a) heading (M) immediately on departure and (b) heading (M) immediately before arrival at B. Given, sin 51° = 0.8.
Answer: (a) 273(M); (b) 270(M). (Hint: the aircraft is flying a great circle path.)

Distance along a parallel or departure

Because the meridians converge, it is clear that change of longitude is only a measure of distance on the Equator, the only parallel which is a Great Circle. The change of longitude on the Equator from 15°W to 20°W, say, is 300 nm but a similar change of longitude at 70°N is nothing like that distance. But there is a very definite relationship.

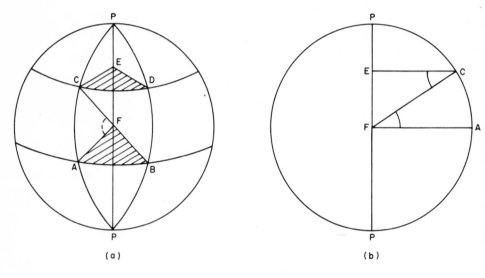

Fig. 3.8

In fig. 3.8(a) AB is the arc of the Equator = change of longitude.

CD is the arc of parallel, of which the distance is required.

(It will be noticed that the distance CD becomes progressively smaller nearer to the Pole, but the change of longitude does not alter).

F is the centre of the Earth; PP the Earth's axis.

∠AFC is the angle of the latitude CD

∵ Triangles ECD and ABF are parallel and equiangular

$$\frac{CD}{AB} = \frac{CE}{AF} \tag{1}$$

In fig. 3.8(b)

$$\frac{CE}{CF} = \text{Cos latitude}$$

∴ CE = CF cos Lat

$$= \text{AF Cos Lat (CF and AF being radii of the Earth)} \tag{2}$$

Substituting (2) in (1)

$$\frac{CD}{AB} = \frac{AF \cos lat}{AF}$$

or, CD = AB cos lat

Thus, distance CD = change of longitude x cosine latitude

The distance made good along a parallel (that is, in an East/West direction) is called <u>Departure</u>.

If the distance and latitude are given, by rearranging the formula Departure = Ch long x Cos lat, we can find the change of longitude.

Dep = Ch long x cos lat

$$\therefore \text{Ch long} = \frac{\text{Dep}}{\text{Cos lat}}$$

= Dep x sec lat

Or, given dep and ch long, latitude could be found:

$$\text{Cos lat} = \frac{\text{Dep}}{\text{Ch long}}$$

Above three variations of the formula are summarised below for convenience:

Departure = Ch long x Cos lat	(1)
Ch long = Dep x sec lat	(2)
Cos lat = $\dfrac{\text{Dep}}{\text{Ch long}}$	(3)

Examples

1. What is the rhumb line distance between A (53°23'N 01°19'W) and B (53°23'N 07°47'W)?

Departure = Ch long x cos lat
= 06°28' x cos 53°23'
= 388' x cos 53°23'
= 231.4 nm

	No.	Log
	388	2.5888
Log cos 53°23'		1.7756
		2.3644
Antilog =		231.4

2. An aircraft takes off from 40°20'N 178°38'E and flies an RL track of 090°. What is its longitude after it has travelled 219 nm?

Ch long = Dep x sec lat
= 219 x sec 40°20'
= 287.3'
= 4°47'
∴ new longitude = 176°35'W

	No.	Log
	219	2.3404
Log sec 40°20'		0.1179
		2.4583
Antilog =		287.3

3. After flying 448 nm along a parallel of latitude an aircraft changes its longitude by 8°21'. What is the latitude?

$$\text{Cos lat} = \frac{\text{Dep}}{\text{Ch long (in minutes)}}$$

$$= \frac{448}{501}$$

= 26°35'N or S

	No.	Log
	448	2.6513
	501	2.6998
		1.9515 = 26°35' from
		log cosine tables.

The following practice problems are graded to become progressively involved. Students taking the CPL exam are normally given the appropriate trig. function values with the questions; ATPL students extract the values from the four-figure log tables issued in the exam. We are following the same pattern.

A reminder before you start: departure is in nm, therefore if you are solving for distance between two places, first convert ch long from degrees to nm by multiplying it by 60. Similarly, when you are looking for ch long, the departure formula will give you the answer in nm; therefore, divide the answer by 60 to get ch long.

Practice problems

1.   Two places on the equator are 8° long apart. If the cosine of 0° is 1, what is the distance between the two places?
Answer: 480 nm.

2.   Two places on 30°S are 8° long apart. What is the distance between them? Given, cos 30° = 0.87.
Answer: 417.6 nm.

3.   Two places on the parallel of 60°N are 8° long apart. What is the distance in nautical miles between them? (Cos 60° = 0.5.)
Answer: 240 nm.

4.   Two places are 360 nm apart on the equator. What is the difference of longitude (d long) between them? (Sec 0° = 1.)
Answer: 6°.

5.   Two places, 360 nm apart are on 60°S. What is the d long between the two places? (Sec 60° = 2.)
Answer: 12°.

6.   What is the departure between A (36°N 10°W) and B (36°N 20°W)?
(Cos 36° = 0.8.)
Answer: 480 nm.

7.   Give the distance in nm between A (41°S 120°W) and B (41°S 125°30'W).
(Cos 41° = 0.75.)
Answer: 247$\frac{1}{2}$ nm.

8.   Give the RL distance between A (51°27'N 00°19'E) and B (51°27'N 13°07'W).
(Cos 51°27' = 0.62.)
Answer: 499.7 nm.

9.   An aircraft flying along the parallel of 81°24' changes its longitude by 55°13'. What distance in nm has it flown? (Cos 81°24' = 0.15.)
Answer: 496.9 nm.

10. If the departure is 230.6 nm along the parallel of 46°31', what is the change of longitude? (Sec 46°31' = 1.45.)
Answer: 5°34$\frac{1}{2}$'.

11. A is at 35°S 140°W; B is on the parallel of 35°S and 370 nm due east of A. What is B's longitude? (Sec 35° = 1.22.)
Answer: 132°28$\frac{1}{2}$' (to the nearest $\frac{1}{2}$ min).

12. An aircraft takes off from 47°30'S 178°30'W and flies a rhumb line track of 270° for 224 nm. What is its new position? (Sec 47°30' = 1.48.)
Answer: 47°30'S 175°59'E (to the nearest degree).

13. How long will it take to go round the earth along the parallel of 75°N at the ground speed of 540 kt? (A computer may be used for GS/time conversion, cos 75 = 0.26.)

Answer: 10 h 24 min.

14. A flight is planned from 40°S 110°E to 40°S 102°E. Give the distance in kilometres if the flight is taking place along the rhumb line track. (Cos 40° = 0.766.)

Answer: 680.9 km.

15. What is the rhumb line distance from Calcutta (30°N 90°E) to New Orleans, U.S.A. (30°N 90°W)? (Cos 30° = 0.866.)

Answer: 9 352.8 nm. (Compare this with the GC distance found in the exercise in chapter 2.)

The above problems are appropriate for both the CPL and the ATPL syllabuses. The following additional problems may be attempted by the ATPL students.

Additional problems

1. Departure = 243 nm; ch long = 4°45'; what is the latitude?

Answer: 31°30'N/S.

2. At what latitude would a distance of 1 066 nm cause a change of longitude of 23°59'?

Answer: 42°12'N/S.

3. Two aircraft, A and B, leave position X (51°N 02°E) for Y (56°N 08°W). Aircraft A on the take-off travels due north until the 56°N parallel is intercepted and then follows that parallel to 8°W longitude. Aircraft B on take-off travels a track of 270°(T) until the longitude of 8°W is intercepted and then flies due north to 56°N. Which of the two aircraft flies the shorter distance and by how many nautical miles?

Answer: Aircraft A travels 41 nm (to the nearest nm) less than aircraft B.

4. An aircraft takes off from 48°05'N 01°W and flies a track of 270°(T) for 2 h 15 min. It then alters its track to 000°(T). Later, it alters its track on to 090°(T) and having flown on this track for 1 h 40 min it recrosses the original meridian of 01°W. If its ground speed was constant at 440 kt throughout the flight what was the flight time on the second leg (Tr 000°(T))?

Answer: 1 h 40 min. Hints: (a) find the ch long on 48°05'N leg; (b) with this ch long find the latitude for the third leg; (c) you now have two latitudes to give the northing.

5. A flight of 2 148 km is being made on a track of 270°(T) along the parallel of 69°N. If the LMT of arrival at the destination is the same as the LMT of departure, what is the flight time?

Answer: 3 h 36 min. (Hint: first find the ch long, then relate this to the speed of the sun's travel which is 15°/hr.)

4: Projections General

Now that we understand the various properties of the earth itself, we are ready to discuss the matter of projections. The object of making a map or a chart is to represent the spherical earth, or at least a part of it, on a flat surface, so that we can record and locate the positions on the earth.

The cartographer works from a model earth (the earth reduced in size to the required scale and called the 'reduced earth'), marked out with the graticule, originally with a light source at some given point within the globe which projects the graticule on to a paper. The paper may be wrapped round the globe like a cylinder, or it may sit on the globe in the form of a cone or it might be just a plane surface, tangential to the globe at a given parallel. Finally he enters the details of ground features as required.

When a projection contains only a graticule of latitude and longitude with perhaps very few geographical features, it is called a 'chart'. A map, on the other hand contains both the graticule and an abundance of ground features.

The cartographer's reduced earth contains all the properties present on the earth. Thus, it would possess the following features.

(a) The scale: it would be both correct and constant. It will be correct because the model earth is a reproduction of the earth on a given scale. It will also be constant, that is, anywhere on the reduced earth a one inch span of the dividers will measure the same distance.

(b) The shapes will be correct.

(c) Areas will be shown correctly.

(d) The bearing measurements anywhere on the reduced earth will be identical to the measurements on the earth.

(e) Great circles will be straight lines, just as on the earth.

(f) Meridians and parallels will intersect each other at 90°.

Projections on flat surfaces

Any attempt at reproducing a spherical shape on to a flat surface must produce inevitable distortions. Thus, on a flat surface we must expect distortions of the ideal properties of the reduced earth as mentioned above. For example, on a flat surface the scale can never be simultaneously correct and constant. Further, the scale is only correct (that is, it remains at the selected value) at a point, e.g. polar stereographic projection, or along a line, Mercator or Lambert.

Similarly, correct shapes are never possible on any map: but this is only a technical distinction. In practice, shapes very close to true shapes are produced on certain projections. Correct areas can be shown at the expense of the shapes.

These are only some of the examples of the limitations. But in spite of them, by systematically arranging the graticule it is possible to control the distortions and produce a projection which meets the requirements very satisfactorily.

Classifications of projections
Projections are classified in a variety of ways and there is no need to get depressed if you do not recognise the title of a projection in your atlas. Here are some of the methods of classification.

According to the method of construction
(a) Geometric or perspective. A flat surface, cylinder or a cone is placed conveniently on the surface of the reduced earth and a light source, again conveniently placed (at the centre or at one of the poles of the reduced earth) projects the graticule. An example is the polar stereographic.

(b) Geometric, modified. The basic geometric projections may be improved by modifying them mathematically. This is normally done to achieve a certain property. Mercator and transverse Mercator projections are examples.

(c) Mathematical. Some projections are produced entirely by mathematical methods, although the basic concept might be drawn from the geometric projection. Lambert's orthomorphic projection is an example here.

According to properties
(a) Equal area projections. On these projections areas are shown in the correct proportions. As the spherical surface is flattened out to give equal areas, the scale, shapes and bearings will all be distorted. A square on the reduced earth may appear as a rectangle or even a parallelogram: as long as the area shown is in the correct proportion, its shape does not matter. As aviators, however, we shall not be concerned with these, but it is useful to know that equal area and orthomorphism (correct bearings) cannot exist together. An example of an equal area is the Bonne's projection seen in any atlas.

(b) Orthomorphic projections. On these projections the bearings are represented correctly. It will be appreciated that for the bearings to be correct the shapes must be correct, and on projections of this type the shapes of *small areas* are preserved. We will come back to orthomorphism pretty soon.

(c) Equidistance projections. In this type of projection the cartographers strive to achieve a constant scale as it occurs on the earth. Again, we shall not be concerned with these projections.

According to the orientation of the paper
(a) Cylindrical projections – Mercator.

(b) Conical projections – Lambert.

(c) Zenithal projections: the plane of the projection is tangential to the earth's surface – polar stereographic.

Projections used in aviation
In aviation we generally use maps and charts for two purposes, that is, navigation plotting and map-reading (topographical maps). In each of these we look for the following properties.

(a) *Plotting charts*

    (i) *Orthomorphism*. This is perhaps the primary requirement of a plotting chart. On this chart the bearings measured are correct.

    (ii) *Rhumb lines*. If the rhumb lines on the chart are straight lines, we would be able to fly a constant track. Although not a primary requirement, it would be very convenient.

    (iii) *Distances*. On any chart used for in-flight plotting, we should be able to measure the distances easily.

    (iv) Lastly, the adjacent sheets should fit to give continuity.

The charts in common use are Mercator's, Lambert's and polar stereographic projections.

(b) *Topographical maps*

    (i) Here we are primarily interested in the shapes which should be shown fairly correctly.

    (ii) Scale on the map should be fairly constant and any scale error distributed evenly over the area.

    (iii) Adjacent sheets should fit accurately.

Transverse Mercator is used for producing topos of regions having their greatest extent north–south, e.g. U.K., Malaysia. Lambert's projection with two standard parallels is suitable for regions having their greatest east–west extent, e.g. U.S.A. Regions roughly circular would advantageously employ oblique stereographic projections whereas in the equatorial regions Mercator's projection is ideal.

Orthomorphism

The word 'orthomorphism' is derived from the Greek words 'orthos' which means right, or correct, and 'morphic' meaning form or shape. The adoption of this title is rather unfortunate as it is bound to confuse those of us who are products of the modern comprehensives and not therefore scholars of classical Greek. So, right from the start we would like to make it clear that on an orthomorphic projection, the bearings measured are correct. The shapes are inevitably tied up with bearing and we will talk about these when discussing individual projections. The American equivalent of the word orthomorphism is 'conformal'. The following considerations go into the construction of an orthomorphic chart.

As we said earlier, as soon as we commence the process of flattening a spherical surface, distortion of shapes takes place, accompanied by inherent distortion of the scale. On the reduced earth the scale is correct. On a flat chart, the distortion must cause variation of the scale. Accepting the limitation that the scale varies from point to point, it is possible to adjust this variation so that on the resultant graticule the scale is the same (though not necessarily correct, that is, equivalent to the reduced earth) along the meridian and parallel through any given point. The scale will have varied at the point next to it, but here again, for a very short distance round it, it will be the same. In other words, the scale may vary from point to point, but the scale around a point remains the same for short distances.

To achieve orthomorphism two requirements must be met. One is that if the chart is to be orthomorphic, the scale at any point measured over a very short distance is the same in all directions, although it may vary from point to point. The validity of this requirement is illustrated in fig. 4.1.

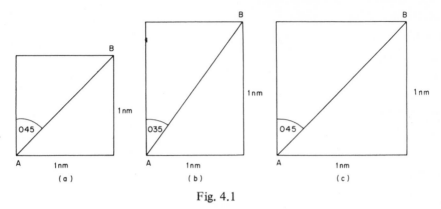

Fig. 4.1

Fig. 4.1(a) represents a square to the scale of $1'' \equiv 1$ nm on the reduced earth. This means that $1''$ on the reduced earth measures 1 nm on the earth. On this square the bearing of B from A is 045°.

Suppose that on a certain projection this square appears as a rectangle as in fig. 4.1(b). The east–west scale on this projection is $1'' \equiv 1$ nm. This scale is the same as on the reduced earth and is therefore correct. But 1 nm in the north–south direction is represented by $2''$ on this projection and consequently the bearing of B from A is wrong. The scale through point A is not the same in all directions and therefore the chart is not orthomorphic.

One yet another projection, fig. 4.1(c), the east–west scale is $2''$ to 1 nm. This is not correct scale, but the scale north–south is also $2''$ to 1 nm. Thus, the scale from point A is the same in all directions and the bearing of B as measured at A is again correct.

It must be emphasised that the above illustration is a very much exaggerated view; the points we are talking about are very close to each other.

The second requirement of orthomorphism is that the meridians and parallels on the projection must intersect each other at right angles, just as they do on the earth. The meridians and parallels on the earth define the north/south and east/west directions and these directions must be reproduced on the chart for the correct measurements. Also, when the meridians and parallels are arranged to intersect at right angles, the shape of a small area about that point is correct which means that the bearings must be correct. It must be noted that the meridians and parallels need not be straight lines in order to intersect at right angles, see fig. 4.2.

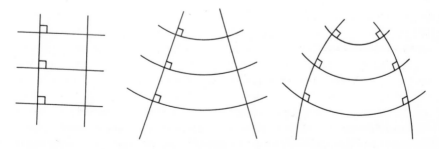

Fig. 4.2

To summarise the two requirements of orthomorphism:

(a) At any given point on the chart the scale must be the same in all directions, although it may vary from point to point.

(b) The meridians and parallels on the chart must cut each other at right angles.

Scales

The scale is the ratio of a distance measured on the chart to the corresponding distance on the earth's surface. It is given in the formula

$$\text{scale} = \frac{\text{chart length (CL)}}{\text{earth distance (ED)}}$$

(both CL and ED must be expressed in the same units).

On the reduced earth the scale is constant and correct everywhere. Thus, if the radius of the model globe was, say, 500 inches, its scale as a representative fraction would be

$$\frac{500}{250\,000\,000} = 1 : 500\,000.$$

This would mean that $1''$ of the model globe would be equal to 500 000 inches on the earth.

For projections where the scale may be varying from point to point, it may not be possible to give a scale which is applicable to the whole coverage. You will find that on some projections (polar stereographic) the scale is given for a point, or a particular parallel (Mercator and Lambert). The maps and charts on which the scale error is extremely small are commonly called 'constant scale' maps/charts. On these charts distances may be measured directly with a graduated rule. The scale itself may be represented in a variety of ways, the representative fraction being the most commonly used method in aviation.

Representative fraction.   This expresses the ratio of a unit of length on the map to its corresponding number of similar units on the earth, e.g. $\frac{1}{1\,000\,000}$. This may also be written in another form, 1:1 000 000. In either case the scale tells us that a distance of 1 inch on the chart represents a distance of 1 000 000 inches on the ground. It also tells us that 1 cm on the chart equals 1 000 000 cm on the ground. In other words, the scale expressed as above has no inherent unit of its own; you provide a convenient unit and the result will be the same provided the same unit is used in both denominator and numerator.

Another point we need to note is that since we are dealing with a fraction of 1, the larger the figure in the denominator the smaller the scale we will have. The scale 1:250 000 is larger than a scale of 1 : 5000 000. On large-scale charts comparatively less ground distance is covered and consequently more ground details can be inserted.

Graduated scale line.   These show the actual lengths on the map corresponding to various distances on the earth. A graduated scale line is found on charts which are considered to have a constant scale throughout the sheet. On some charts (Lambert's six million) more than one such line is found. For more accurate distance measurement a variable scale may be used, but this is beyond the scope of our present studies.

Statement in words.   This method, generally used on Ordnance Survey maps gives the corresponding values of two different units of length, one on the map and the other on the earth, e.g. 'one inch to one mile'. By common usage, large-scale maps are referred to by the scale of the map — 'one inch map'. This would imply the scale of one inch to one mile. Similarly, small-scale maps are referred to by the earth distance represented by one inch — 'a ten mile map'.

It is worth mentioning that one in quarter million scale (1: 250 000) is approximately $\frac{1}{4}$ inch to the mile or 1 cm to 2.5 km.

In addition to the above it must be noted that the meridians, being semi great circles, provide a natural nautical mile scale. One degree change of latitude along a meridian measures 60 nm on the earth and therefore the meridians are usually graduated in nautical miles. These graduations are very useful for measuring distances.

Scale factor

We stated earlier that one of the steps in the construction of a projection is to reduce the earth to a chosen scale. It is this scale that is generally printed on the chart. We also stated that on any flat chart the scale is bound to depart from the chosen scale of the reduced earth. In detailed study of the scales, it is sometimes more convenient to calculate how much the scale differs from the chosen scale at a given place, rather than the actual scale there. This difference from the reduced earth (RE) scale is called the 'scale factor', and it is expressed as a ratio of the chart length to the reduced earth length

$$\text{scale factor} = \frac{\text{chart length}}{\text{reduced earth length}}.$$

It can be seen from the above relationship that the scale factor will be 1 wherever the scale on the chart is identical to the scale of the reduced earth. Thus, when the ratio is not exactly 1 the difference between it and 1 is the amount the scale differs from the reduced earth scale. For example, if the scale factor at a point has a value of 1.0123, this means that a measurement of 1 unit at this point will be in error by 0.0123 and the percentage error will be 1.23.

Alternatively, knowing the scale factor at a point, we can work out the actual scale at that point. If the reduced earth scale is 1:1 000 000 and the scale factor at point A is 1.2 then the scale at A is equal to

$$\frac{1}{1\,000\,000} \times 1.2 = 1 : 833\,333$$

and note that the scale at point A is larger than the reduced earth scale.

Scale problems

Problems involving scales are often set in the exams. We give you here a few worked examples followed by plenty of practice problems. You may find the following conversion factors (in addition to the ones given in an earlier chapter) useful in working out your scale problems. Log tables are not available in the CPL exams and there is a heavy penalty for making five from two plus two.

    1 nm = 72 960 inches
    1 sm = 63 360 inches
    1 km = 100 000 cm
    1 inch = 2.54 cm

Worked examples

1.   How many nm to an inch are represented by a scale of $1:2\,500\,000$?

$$1 \text{ inch represents } 2\,500\,000 \text{ inches or } \frac{2\,500\,000}{6\,080 \times 12}\text{nm}$$

$$= 34.26 \text{ nm.}$$

2.   You have a chart, scale 4 inches to 1 statute mile. Express this as a representative fraction.

$$\text{Scale} = \frac{CL}{ED} = \frac{4 \text{ inches}}{1 \times 5\,280 \times 12}$$

$$= 1:15\,840$$

3.   If 100 nm are represented by a line 7.9 inches long on a chart, what is the length of a line representing 50 km?

$$7.9 \text{ in} = 100 \text{ nm} = 185 \text{ km}$$

$$\therefore 50 \text{ km} = \frac{50 \times 7.9}{185}$$

$$= 2.14 \text{ inches}$$

4.   If the scale is $1:250\,000$ what is the distance on the chart between $32°11'$N $06°47'$E and $30°33'$N $06°47'$E?

Since both positions are situated on the same meridian the distance between the two in nautical miles is the arc of the meridian in number of minutes ($1° = 60$ minutes) intercepted between the two, viz, change of latitude in minutes.

Thus the distance in nm = $32°11'$
$$-30°33'$$
$$1°38' \text{ or } 98 \text{ nm.}$$

Now we proceed to solve the problem in the usual way — 1 inch on the chart represents 250 000 inches on the earth, or

$$\frac{250\,000}{72\,960} \text{ nm on the earth.}$$

$$\therefore 98 \text{ nm will be represented by } \frac{98 \times 72\,960}{250\,000} \text{ inches}$$

$$= 28.6 \text{ inches.}$$

Practice problems

1.   Which of the following two charts has a larger scale? (a) $1:1\,000\,000$; (b) $1:250\,000$.

Answer: Chart (b).

2.   Which is the larger scale, $1:1\,303\,000$ or 17.8 nm to one inch?

Answer: 17.8 nm to one inch.

3.   If the scale is $1:500\,000$, how many statute miles on the ground are represented by one inch on the map?

Answer: 7.89 sm.

4.   A Mercator's chart has a scale of $1:1\,461\,000$ at 50°N. How many nautical miles are represented by one inch at 50°N on this chart?

Answer: 20.02 nm.

5.   If on a chart, 30 statute miles are represented by 30 centimetres, what is the scale of the chart?

Answer: 1:160 934.

6.   On a chart, scale 1:1 000 000, how many inches does a ground distance of 286 kilometres measure?

Answer: 11.26 inches.

7.   A line 6 inches long drawn on a chart measures 105 nm. What is the scale of the chart?

Answer: 1:1 276 800.

8.   On a chart having a scale of 1:2 000 000, how many kilometres are represented by a line 3.7 inches long?

Answer: 187.96 km.

9.   On a chart having a scale of 1:250 000, at what distance in inches would two positions A (20°33'N 150°08'W) and B (21°37'N 150°08'W) appear?

Answer: 18.68 inches.

10. On a chart, a line 8 inches long represents 200 statute miles. What is the length of a line in centimetres representing 326 kilometres?

Answer: 20.58 cm.

11. At what distance would two pinpoints taken at a 20 minute interval appear on a chart, scale 1:1 000 000 if the ground speed was 180 kt?

Answer: 4.38 inches.

12. If the scale of a chart is 1:6 000 000, how many kilometres are represented by a line 5.2 inches long?

Answer: 792.48 km.

13. Position A on latitude 46°44'S is due north of position B which is on latitude 49°39'S. If the distance between the two on a constant scale chart is 6.01 cm, what is the scale of the chart?

Answer: 1:5 396 126.

14. An aircraft takes 15 min 12 s to cover the distance 6.6 cm between A and B on a chart having a scale of 1:2 000 000. Calculate the aircraft's ground speed.

Answer: 281.6 kt.

15. The distance flown by an aircraft in 40 seconds at a ground speed of 480 kt is shown on a chart by a straight line, 1.4 inches long. Give the scale of the chart in centimetres to kilometres.

Answer: 1 cm to 2.76 km.

Chart symbols and relief

You learn all this from the maps and charts you come across in your flying activities from the word go: Civil Aviation Department Aeronautical Information Circular entitled Aeronautical Charts provides comprehensive information on various maps and charts available and the ICAO approved symbols. On radio navigation charts, the legend is available. Many maps print the key on the side of the map. Thus, as the psychologists say, this sort of knowledge is acquired 'by use and wont'. The only word of warning is spot heights: these are indicated by a dot with a figure by the side Before using any chart, look around the margin to check if spot heights are given in feet or metres. 123 metres, for example, is about 404 feet — quite a difference.

Relief on the topographic charts is usually shown by contours with a spot height. The nearer the contours the steeper the hills. Layer tinting is another method of showing relief. The depth of the colour as keyed at the side of the map indicates the gradient,

again with a spot height to show the highest spot in the area. This boy scout stuff –
which just comes naturally after the initial reading and use of various charts – should
not be taken lightly by armchair leisurely browsing at first.

5: Mercator's Projection

Gerhard Kramer, a Flemish mathematician, was fifty-seven years old when he published his cylindrical orthomorphic projection in 1569. It was modish to have a Latin surname to be distinguished from the unscholarly plebeians, hence Mercator. The copies of his chart still exist but he left no clue as to how he calculated the spacing of the graticule. Thirty years later Edward Wright of Cambridge published the tables for the construction of the projection in his book *Certaine Errors in Navigation*. Accurate tables, called meridional parts tables did not become available until a hundred years later when calculus opened up a new dimension in mathematics. Perhaps the most used projection in the world, Mercator was particularly popular in Britain because, they say, the British Empire on this projection looked bigger than the rest of the world. We will now consider the three stages in its development.

Geometric cylindrical

A roll of paper in the form of a cylinder is placed round the reduced earth, its axis being coincident with the axis of the generating globe. Thus, the cylinder is tangential to the reduced earth at the equator, see fig. 5.1. A light source placed at the centre of the generating globe casts shadows of the graticule on to the developable surface. The resultant graticule is rectangular in shape.

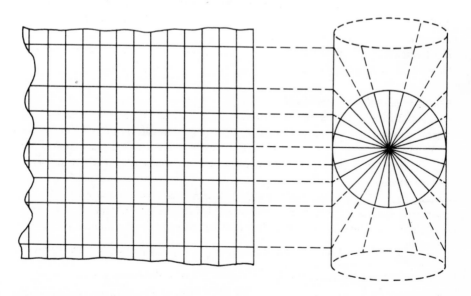

Fig. 5.1

The primary feature of all normal cylindricals is the property of straight and parallel meridians, equally spaced, at right angles to the equator. ('Normal' cylinder means a cylinder having its axis coincident with the axis of the reduced earth.) Thus, the first step in drawing up any normal cylinder is to rule in the meridians against the equator or any other parallel.

The next step is to insert the rest of the parallels. The spacing of these parallels give the individuality to the cylinder. However, note from fig. 5.1 that all parallels are of equal length, the length of any parallel on the cylinder being the same as the reduced earth, that is, $2\pi R$.

The scale of such a geometrical projection must be worked out separately in east—west and north—south directions.

East—West scale

$$\text{Scale} = \frac{\text{CL}}{\text{ED}}$$

$$= \frac{\text{length of any parallel } \phi \text{ on the projection}}{\text{length of the same parallel } \phi \text{ on the RE}}$$

$$= \frac{2\pi R}{2\pi R \cos \phi}$$

$$= \frac{1}{\cos \phi} = \sec \phi$$

North—south scale
The scale factor in the north—south direction on this projection can be calculated by calculus, and it is $\sec^2 \phi$.

Properties.   The great difference in the rate of change of the scale in the two directions as we saw above renders the distance measurement a highly complex business. In addition, the scale variation from a point being different in different directions, the chart is not orthomorphic. These two major disadvantages rule out its use for plotting or as a topographical map.

Simple (equidistant) cylindrical
This is the second stage in the development of a cylinder. Whereas the ancient mariners were not to be fooled by the geometric cylindrical, they loved this one as it looked so 'earth-like'. It was widely used by the navigators before the arrival of the real science of map projection and Mercator, and in consequence discovered many remote islands.
Construction.   This is a straightforward mathematical projection in which the calculated distances are plotted on the map sheet to give the graticule. The equator is marked in as a straight line. It is then divided correctly for the meridian spacing $\left(\frac{\text{ch long}}{360} \times 2\pi R\right)$, and the meridians drawn in as straight and parallel lines at right angles to the equator (fig. 5.2). The meridians are then divided to give the parallel spacing at *correct* reduced earth scale $\left(\frac{\text{ch lat}}{360} \times 2\pi R\right)$ and the parallels drawn in. The

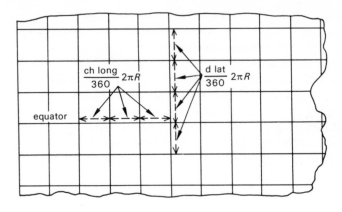

Fig. 5.2

scale in the east–west direction on such a chart is sec ϕ, found in the same way as for the geometric. The scale factor in the north–south direction is obviously 1, since the scale has been made constant and correct by construction.

Properties.   Distance measurement in directions other than due north–south or due east–west is complicated. It is not orthomorphic as it does not meet one of the two requirements of orthomorphism. The chart is no longer in use.

Orthomorphic cylindrical or Mercator's projection

This is the one we are interested in most. In this projection the spacing of the parallels is deliberately chosen in such a way that with increase of latitude, the rate of expansion of scale in the north–south direction proceeds at the same rate as the expansion of scale in the east–west direction (which is secant of latitude). See fig. 5.3. Thus, at any one point the scale will have expanded by the same amount, will have reached the same value and will in fact be the same in all directions.

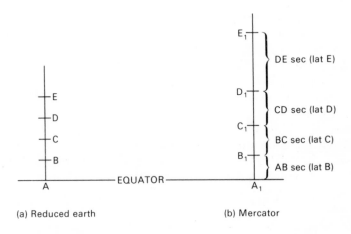

Fig. 5.3

As in other normal cylindricals the scale factor in the east—west direction is sec ϕ because converging meridians have become straight and parallel. Thus, the rate of expansion required in the north—south direction will be such as to give the scale factor of sec ϕ at any latitude. Putting this the other way, the meridians will be stretched as they depart from the equator at a rate proportional to the secant of the angle of latitude. The actual process of calculating the spacing involves the use of calculus and meridional parts tables are available. The construction of a Mercator is not in your syllabus until you take the Flight Navigator's licence. However, for the ambitious and the interested, a specimen construction is shown in appendix 1. Our primary interest in the study of Mercator is in understanding its properties.

Properties
Orthomorphism. Yes, the chart is orthomorphic by construction. It meets both the requirements of orthomorphism.
Where on the chart is the scale correct? The cylinder is tangential to the equator of the reduced earth. Therefore, it is only here that the two lengths are identical, to give the correct scale. Thus the answer is, only at the equator.
How does the scale vary? Away from the equator, the scale varies as secant of the latitude. The same chart length at 30°N will measure a larger distance than at 60°N. For this reason the scale of a Mercator is given as a representative fraction for a stated latitude, e.g. 1:1 000 000 at 51°N.
Convergency. Convergency is defined as the inclination between two meridians. On Mercator the meridians are straight lines parallel to each other. Consequently, the angle of inclination is 0° throughout. Thus, the convergency on the chart is *constant*, and because it is constant, it cannot be correct anywhere on the chart except at the equator where the earth convergency is also 0°.
Great circles. Great circles on the earth cut meridians at different angles. These must be curves on this chart where the meridians are parallel to each other. They curve *concave* to the equator. However, the equator and all the meridians together with their anti-meridians are great circles by definition, and they appear as straight lines. (They are rhumb lines as well.)
Rhumb lines. All straight lines drawn on this chart are rhumb lines because they cut all the meridians at the same angle.
Shapes. As the projection is orthomorphic, shapes of small areas are not distorted. The shapes of large areas in high latitudes where the scale expansion becomes very rapid are considerably distorted. An example is the shape of Greenland on this projection. A circle at the equator will appear egg-shaped at, say, 60°N. This effect of the scale expansion in high latitudes stretches its property of orthomorphism until it finally cannot be used for plotting.
Areas. Mercator's projection is not an equal area projection. Areas are greatly exaggerated in the high latitudes. South America which is ten times the size of Greenland actually appears smaller than Greenland on this projection.
Fits. The sheets with the same meridian spacing will fit accurately in north—south and east—west.

Mercator as a plotting chart
Mercator meets all the requirements we would look for in a plotting chart. It is ortho-

morphic, rhumb lines are straight lines and it has a satisfactory scale in the lower and intermediate latitudes.

Limitations of Mercator
Due to the expanding scale, the chart cannot be used for plotting purposes in high latitudes, say, above 70°/75°N/S. The rapidly expanding scale makes accurate distance measurements difficult, and eventually the bearings are distorted. The poles cannot be produced on the chart.

Disadvantages of the Mercator
1.   The major one we have already met — radio bearings are great circle bearings which must be converted to rhumb lines before plotting.
2.   On a long drag, the rhumb line can add appreciably to the distance. It is then usual to plot the latitude and longitude of a series of points which constitute the great circle and join these points by rhumb lines forming a composite RL track. This was common over the Atlantic and Pacific; nowadays, there is no such thing as a long drag with increasing aircraft speeds and the Lambert's projection can be used giving distances very close to great circle distances.
3.   The chart cannot be used in polar regions.

Measurement of distances
Since the scale on the chart varies from place to place, care must be exercised in taking distance measurements. Always use the latitude scale (up a meridian) using a convenient section which straddles the middle of the line you are measuring. Keep that section for the whole of the line. For large distances, say in excess of 300 nm, section out the line to be measured and take a mean latitude cover for each section. Then measure each section separately and add up for total distance. Short distances readily spanned by the divider can be measured in one attempt over the mid-latitude of the line concerned.

To form a local graticule
A local graticule of the Mercator can be constructed as follows:
    Say you wish to produce a graticule covering 55°N to 57°N, 100°W to 105°W.
1.   Choose a scale for longitude, say 1 in ≡ 1° longitude.
2.   Draw a base line AB and mark off longitudes at a distance of 1 in per degree.
3.   Construct the meridians as parallel lines perpendicular to the base line.
4.   From the point of origin A draw an angle *from the base line* equal to the mean of the required parallels (see fig. 5.4), and let this intersect the adjacent meridian at C.
5.   From A, with radius AC, draw an arc CD.
6.   At D, draw the line of latitude, parallel to the base line.
7.   Repeat the process as required.
    The points to watch are that the angle is from the base line and not from north on the protractor; and relate the degrees of longitude to the degrees of latitude — one degree spacing in our sketch, but such a scale for 5° spacing of longitude for example must refer to a similar 5° spacing of latitude.

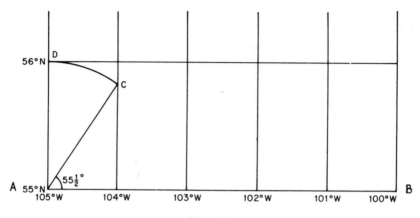

Fig. 5.4

Plotting radio position lines

As all radio bearings are great circles, they must be converted to rhumb lines before plotting on a Mercator. This is readily done by applying conversion angle either from the formula CA = ½ ch long x sin mean lat, or from the abac scale printed and explained in the left-hand corner of the chart, and demonstrated in *Ground Studies for Pilots* Volume 2, 'Plotting and Flight Planning'. Apply to the nearest whole degree, and bear in mind that the RL always appears towards the equator. A thumbnail sketch is always useful.

Conversion angle is applied *where the work is done*.

Example

You are in DR position 30°N 20°W and receive a QTE from a station in 35°N 23°W of 123°. The work of measurement has been done at the station; the conversion angle

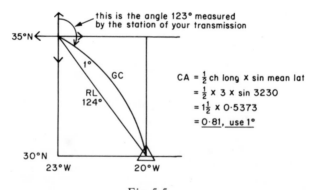

Fig. 5.5

must then be applied to this measurement: the GC bearing is 123°, add CA of 1°, and plot this RL of 124° from the station.

Now if the circumstances had been similar to the one quoted above, but you decided to measure the bearing of the station yourself from the aircraft, then what you measured is the subject for conversion; in other words, you have done the work. Say your heading is 350°(T), and you took a relative bearing of the station on the radio compass of 315°.

Relative bearing	315 GC
Add heading true	350
	665
	−360
True bearing	305 GC − this is the work done
Conversion angle	− 1 to bring it nearer to equator
True bearing	304 RL
	−180 to get to the station for plot
Plot	124°

Consider this example in the southern hemisphere. QDM is 050°, variation at the station 5°E, CA = 4°.

Apply the station's variation to the QDM to give a heading to steer with zero wind of 055°(T)

$$055$$
$$+180 \text{ to get to the station}$$
$$235 \text{ the actual work done}$$
$$CA + \quad 4 \text{ to bring it nearer to equator}$$
$$\text{Plot } 239°$$

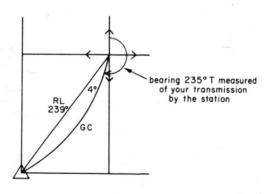

bearing 235° T measured of your transmission by the station

RL 239°

4°

GC

Fig. 5.6

Practice paper: Mercator North

Note: use CA throughout as 1° when its value is 0.5 or more.

Construct a Mercator graticule from 20°N to 25°N, 10°W to 15°W using the scale of 1 in ≡ 1° longitude. Then answer the following questions based on it:

1.   Estimate RL track angle and distance from A (20°N 14°W) to B (24°30′N 10°30′W).

2.   Would RL track angle from 25°N 14°W to 29°30′N 10°30′W be more, same or less than answer in 1 above? Why?

3. What is the scale of your chart at the equator?
4. What is GC bearing of (i) A from B; (ii) B from A?
5. Is the convergency of the chart correct?
6. Is the convergency of the chart constant?
7. Where is the convergency of a Mercator chart constant?
8. Where is the convergency of a Mercator chart correct?
9. Why is a Mercator chart of little or no navigational value in high latitudes?
10. Give the track from F (22°30′N 15°W) to K (22°30′N 10°W) which will give the shortest distance.
11. Is your chart orthomorphic? Why?
12. Measure on your chart the GC bearing of M (25°N 13°W) from N (20°N 13°W).
13. An aircraft in DR position 20°20′N 14°50′W on heading 050°(T) obtains a relative bearing of 355° from station at 23°12′N 10°W. What would you plot?
14. An aircraft in DR position 24°30′N 15°W gets QDM 126° from station at 20°N 10°W. Station variation 9°E. What would you plot?

Answers:

1. 036°; 329 nm.
2. Less — expanding scale (meridians converge on the earth).
3. 1 in = 60 nm at the equator; therefore scale is 1:4 377 600.
4. (i) 217°; (ii) 035°.
5. No.
6. Yes.
7. All over the chart.
8. At the equator.
9. Scale expansion.
10. Shortest distance is along GC track which is 089° (initial) 090° (mean) and 091° (final).
11. Yes, two basic requirements of orthomorphism are met.
12. 000°.
13. 226°.
14. 314°.

Practice paper: Mercator South

Note: use CA to the nearest whole degree.
Construct a Mercator graticule from 20°S to 50°S, 30°E to 60°E (showing the meridians and parallels at 5° intervals) using scale 1 in ≡ 5° longitude, and answer the following questions.

1. Two lines are drawn from 35°E to 55°E, the first at 22°S, the second at 47°S.
    (i) Which represents the longer distance?
    (ii) Which has the smaller scale?
    (iii) What is the scale of this chart at the equator?
2. How many km are expressed by 1 cm at the equator?
3. Point A is at 50°S 60°E; point B is at 25°S 35°E.
    (i) What is RL track and distance from A to B?
    (ii) Would RL track angle from 75°S 60°E to 50°S 35°E be greater or lesser than (i) above? Give a reason.
    (iii) What is the initial GC track from (a) A to B; (b) B to A?

(iv) What is the mean GC track from A to B?
4.   What track to nearest degree will give the shortest distance between
(i) 47°S 30°E and 47°S 55°E? (ii) 47°S 30°E and 20°S 30°E?
5.   Express the scale of this chart as a representative fraction.
6.   An aircraft in DR position 40°S 55°E receives QDM 321° (station variation
18°W) from X (30°S 30°E). What would you plot?
7.   An aircraft in DR position 49°S 46°E on heading 090°(T) gets radio compass
bearing (relative) of 268°, QC 0, from station at 25°S 45°E. What would you plot?
Answers:
1.   (i) Line drawn at 22°S.
     (ii) 22°S.
     (iii) 1: 21 888 000.
2.   218.9 km.
3.   (i) 322°; 1 950 nm.
     (ii) Greater – expanding latitude scale.
     (iii) (a) 314°; (b) 150°.
     (iv) 322° – GC and RL are same at mid-meridian.
4.   (i) Initial 099°; mean 090°; final 081°.
     (ii) 000°.
5.   1:17 929 000. The scale is found in two ways – (i) measure the distance at mid-
position on the chart (along a meridian) by dividers open one inch. Convert this
distance as representative fraction. (ii) By calculation. The latter method is more
accurate and is explained in the following pages.
6.   116°.
7.   178°.

Scale problems on Mercator

In tackling any problems on Mercator in which scale is involved we have only to
remember that the scale varies as secant of latitude.

1.   If the scale at equator is 1: 608 776, what is the scale at 56°N?
The scale expands at the rate of secant of latitude, therefore
scale at 56°N = scale at equator x sec 56°

$$= \frac{1}{608\,776} \times \sec 56°$$

As the numerator of the RF must be 1, the equation is rewritten:

$$= \frac{1}{608\,776 \times \cos 56}$$
$$= 1:340\,427$$

It is noted here, by the way, that the denominator at 56°N is a smaller figure than
at the equator. This indicates scale expansion; 1 inch at 56°N will measure lesser
earth distance.
2.   Scale at 56°N is 1:1 000 000. What is the scale at 46°N?

$$\text{Scale at equator} = \frac{1 \times \cos 56°}{1\,000\,000}$$

$$= \frac{1}{1\,000\,000 \times \sec 56}$$

$$\therefore \text{Scale at } 46°N = \frac{1}{1\,000\,000 \times \sec 56 \times \cos 46}$$

$$= 1:1\,242\,332$$

3.   Scale at 50°N is 1:608 000. What is the scale at 60N?

$$\text{Scale at } 60°N = \frac{1}{608\,000 \times \sec 50 \times \cos 60}$$

$$= 1:472\,930$$

4.   An aircraft flies from CLINTON, 43°37′N 81°30′W to NIAGARA, 43°07′N 79°00′W. The distance between the meridians of CLINTON and NIAGARA on a Mercator chart is 15 in. What is the scale of the chart at 45°N?

   Change of longitude = 2°30′ represented by 15 in on the chart.

$$\therefore 1° \text{ ch long} = 6 \text{ in}$$

and since the meridians on a Mercator are straight lines parallel to each other

   1° at the equator   = 6 in; or,

   60 nm at the equator = 6 in

$$\therefore 10 \text{ nm at equator} = 1 \text{ in}$$

and at 45°N the scale of the chart is

$$= \frac{1}{10 \times 6\,080 \times 12 \times \cos 45}$$

$$= 1:515\,900$$

Practice problems: Mercator's projection

Exercise 1 (CPL and ATPL)

1.   If the scale at the equator is 1:6 000 000, what is the scale at 45°S?
(Sec 45° = 1.41, cos 45° = 0.71.)
Answer: 1:4 260 000.

2.   If the scale at the equator is 1:1 427 000, what is the scale at 58°N?
(Sec 58° = 1.88, cos 58° = 0.53.)
Answer: 1:756 310.

3.   If the scale at 60°S is 1:3 000 000, what is the scale at the equator?
(Sec 60° = 2.0, cos 60° = 0.5.)
Answer: 1:6 000 000.

4.   If the scale at 30°N is 1:2 450 000, what is the scale at the equator?
(Sec 30° = 1.15, cos 30° = 0.87.)
Answer: 1:2 817 500.

5.   If the scale at 30°N is 1:3 500 000, what is the scale at 40°N? (Cos 30° = 0.87, sec 30° = 1.15, cos 40° = 0.77, sec 40° = 1.30.)
Answer: 1:3 099 250.

6.   If the scale at 60°N is 1:2 000 000, what is the scale at 40°N? (Cos 60° = 0.5, sec 60° = 2, cos 40° = 0.77, sec 40° = 1.30.)
Answer: 1:3 080 000.

7.   The scale at the equator is 1:4 000 000. The distance as measured on the chart by a straight line between A (60°N 70°W) and B (60°N 80°W) is 9.2 inches. (Cos 60° = 0.5, sec 60° = 2.0.)
   (a) What is the scale at 60°N?
   (b) What is the chart length in centimetres along the equator representing 60 nm?

(c) What is the earth distance in kilometres between A and B?

Answers: (a) 1:2 000 000; (b) 2.34 cm; (c) 555 km.

8.   On a Mercator, two meridians one degree apart are spaced at a distance of 7.62 cm. What is the scale on this chart at 48°N? (Sec 48° = 1.49, cos 48° = 0.67.)

Answer: 1:977 664.

9.   On a Mercator the straight line distance between two positions A (40°S 08°E) and B (40°S 08°W) is 88 cm. What is the chart length in millimetres between 12°W and 13°W at the equator?

Answer: 55 nm.

10. At 60°N a distance of 30 cm on the chart measures a ground distance of 390 nm. What is the scale of the chart at the equator?

Answer: 1:4 810 000. (Hint: to avoid awkward figures, work in kilometres.)

Exercise 2 (ATPL: use four-figure log tables)

1.   If the scale at 40°N is 500 000, where is the scale of 1:450 500?

Answer: 46°21′N/S.

2.   If the scale at 42°N is 1:1 000 000, where is the scale of 1:500 000?

Answer: 68°11′N/S.

3.   In problem 2 above, is it possible to have a scale of 1:2 000 000 anywhere on the chart?

Answer: No. (Hint: compare it with the scale at the equator.)

4.   On a Mercator one degree of change of longitude measures 3.03 cm. In which latitude is the scale 1:2 000 000?

Answer: 56°59′N/S.

5.   On a Mercator the spacing between the meridians of 160°W and 170°W is 26 cm. What is the scale at 56°N?

Answer: 1:2 392 000.

6.   On a Mercator the distance between meridians 172°W and 159°W is 9 inches. At what latitude would you find the scale of 1:3 000 000?

Answer: 61°41′N/S.

7.   If the scale at 60°N is 1:2 000 000, where is the scale of 1:1 000 000?

Answer: 75°31′N.

8.   If the scale at 57°20′N is 1:1 091 000, what is the meridian spacing, one degree apart, in centimetres?

Answer: 5.5 cm.

9.   The scale on a Mercator at 42°S is 1:500 000. What is the distance in inches between two places A (42°S 112°E) and B (42°S 115°42′E) on the chart?

Answer: 24.1 inches.

10. A Mercator has a scale of 1:1 000 000 at the equator. On this scale, two positions A and B both on 54°N are shown 10 inches apart. What is the difference in longitude between them?

Answer: 2°17.1′.

6: Lambert's Conical Orthomorphic Projection

This projection devised by Johannes Lambert and first published around 1772 is variously known as

    Lambert's conical orthomorphic projection;

    Lambert's 2nd projection;

    Lambert's conformal.

It is an improvement on Mercator's projection: its scale is very nearly constant and great circles are almost straight lines. The difficulty caused by rhumb lines not being straight lines is overcome either by superimposing a square grid on it or by simply flying mean tracks. With the ICAO's blessing this projection is now progressively replacing Mercator as a plotting chart.

    The projection is entirely mathematical and the resulting graticule has some similarity to the geometric conic with two standard parallels. In fact, this projection is based on conic conception and therefore we shall first look into the properties of a simple conic projection.

Simple conic projection

A cone is placed over the reduced earth so that it is tangential to a predetermined parallel of latitude. This parallel is called its 'standard' parallel. A standard parallel on any projection is that parallel which is drawn to the same scale as the scale of this parallel on the reduced earth. A Mercator's standard parallel is the equator.

    In fig. 6.1 a cone EAXBQ is placed on the reduced earth, tangential along the

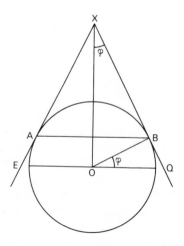

Fig. 6.1

parallel AB, which is its standard parallel. Parallel AB is also the 'parallel of origin' of the chart. Its radius from X, the apex of the cone, is XB and is given in the formula R cot ϕ (where R is the radius of the reduced earth and ϕ is the standard parallel). The formula can be proved quite simply.

AB is the standard parallel which is described by angle OXB, ϕ. OB is the radius of the reduced earth, R. (It can be seen that the lat angle QOB = ∠OXB.)

In triangle XOB

$$\frac{XB}{OB} = \cot \phi; \therefore XB = OB \cot \phi$$

$$= R \cot \phi \text{ (OB is the radius).}$$

Construction
The following steps describe the method of drawing up a simple conic.

(a) Draw the central meridian.

(b) From vertex X draw in the standard parallel as a segment of a circle, radius XB being R cot ϕ.

(c) From the vertex draw the meridians cutting the standard parallel at uniform spacing. This spacing is calculated from $\frac{ch\,long}{360} \times 2\pi R \cos\phi$.

(d) Mark off the positions for placing the remaining parallels along the central meridian. As the length of the meridian is $2\pi R$, the spacing is $\frac{ch\,lat}{360} \times 2\pi R$. Thus, all parallels are at a constant distance from each other just as they are on the reduced earth. Draw in the parallels to complete the graticule.

Properties
Scale. The scale is correct along the standard parallel and all the meridians. It is too large along the other parallels, increasing with distance away from the standard parallel.
Orthomorphism. As all the meridians are drawn as straight lines to the parallels from a common point, they cut each parallel at 90° and one requirement of orthomorphism is met. But as the scale variation from a point is not the same in all directions, the chart is not orthomorphic.
Convergency. As the meridians are drawn as straight lines, the convergency is constant throughout the chart, the convergency factor being called 'the constant of the cone' and expressed by the letter n.

Constant of the cone
What determines the actual value of the convergency on a conical chart? It will be seen in fig. 6.2 that when the cone covering the surface of the reduced earth is unwrapped, the paper containing the graticule is not a full circle but only a sector of a circle. In fig. 6.2, AB is the standard parallel, starting from the Greenwich meridian and going round the earth and back to the Greenwich meridian. But the angle through which it represents the earth's 360° is in fact less than 360° and this is the measure of its convergency.

The ratio of the angle subtended by the arc AB to the length of the same latitude on the earth is called the constant of the cone. Its value varies from 0 for a Mercator

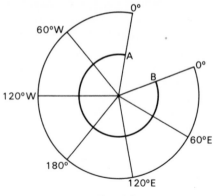

Fig. 6.2

(Mercator may be considered as an extreme case of a cone, having an apex angle of 0°) to the value 1 on polar stereographic (which may be considered the other extremity of the cone, having an angle of 180° at the apex). For a simple cone it will therefore remain between 0 and 1, the actual value being dependent on the chosen standard parallel.

The value of the constant of the cone, n, may be ascertained from a comparison of the length of the standard parallel to the actual length on the earth.

The length of a parallel on the earth is equal to $2\pi R \cos \phi$. As the standard parallel is correct to scale:

$$2\pi R \cot \phi \times n = 2\pi R \cos \phi$$

and
$$n = \frac{2\pi R \cos \phi}{2\pi R \cot \phi}$$

$$= \frac{\cos \phi}{\cot \phi}$$

but
$$\cot \phi = \frac{\cos \phi}{\sin \phi}$$

∴
$$n = \frac{\cos \phi \sin \phi}{\cos \phi}$$

$$= \sin \phi$$

where ϕ is the standard parallel.

Thus, the convergence on this projection is the sine of the standard parallel. For a standard parallel of say 50°N, the chart convergence is sin 50° = 0.7660 and the angular inclination between two meridians 10° apart anywhere on the chart will measure 7.660°.

The geometric conic can be mathematically modified by placing the parallels using the 'orthomorphic' formula instead of placing them at a constant distance. This would produce a strict orthomorphic projection. But in our syllabus we are primarily interested in the orthomorphic based on the two standard parallel conic and we will talk about this now.

Conic with two standard parallels

As the title of the projection suggests, two parallels are drawn correctly to scale.

Consequently, the scale between the two standard parallels contracts and the scale outside the standard parallels expands, see fig. 6.3. The meridians are drawn in as straight lines radiating from the cone's vertex. The spacing of the parallels along the meridians gives true scale along the meridians.

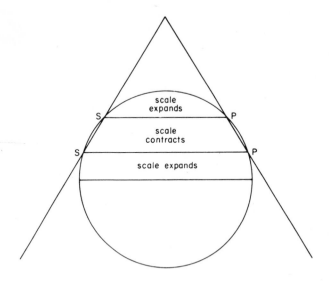

Fig. 6.3

The chart is not orthomorphic as the scale variation requirement is not met. However, since we have two parallels on the projection giving correct scale, the accuracy of the mapped area as well as the coverage itself is better than the simple conic. Countries having an east–west extent rather than a north–south one may be mapped on this projection. Maps of the U.K. (ch lat 10°) have been successfully made on this projection for atlases, however.

Lambert's orthomorphic projection with two standard parallels

We saw earlier that on a conic with two standard parallels there is a scale contraction between the standard parallels and expansion outside them. Thus, the scale along the parallels varies from one parallel to the next whereas the scale along the meridians remains constant. To make the chart orthomorphic, the meridian scale must be modified in such a way that at any point along a meridian the scale is equal to the scale of the parallel passing through that point. Lambert achieved this by use of his orthomorphic formula. For interest only a full construction of the projection is shown in appendix 2. The radius of each parallel is calculated from this formula and drawn in as an arc of a circle from a common centre (the apex). The two standard parallels are then divided to give a uniform meridian spacing using the formula so well known to us by now, $\dfrac{\text{ch long}}{360} \times 2\pi R$, and the meridians are inserted as straight lines. The completed graticule is rectangular in shape except when very close to the equator, see fig. 6.4.

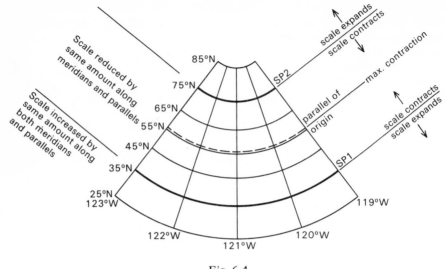

Fig. 6.4

Properties
Orthomorphism. The projection is orthomorphic by construction.
Where is the scale correct? The scale is correct along the two standard parallels. This means that the scale is the same as on the reduced earth.
How does the scale vary? It contracts in between the two standard parallels and expands outside the standard parallels, fig. 6.4.

Thus a span of say one inch on the dividers measuring 5 nm along a standard parallel will measure more than 5 nm inside the two parallels, the maximum distance being measured (approximately) at a point midway between the two. This is the chart's parallel of origin. Similarly, outside the standard parallels it will measure less than 5 nm. In terms of representative fractions, if the scale was given as, say, 1:250 000, this is true for the two standard parallels. The denominator will be greater than 250 000 inside the two standard parallels and smaller to the outside.

Outside the standard parallels, the expansion towards the poles is not in the same proportion as towards the equator. The rate of expansion towards the poles is slightly higher than towards the equator. However, this distinction is of technical interest only. In fact, on some large-scale charts e.g. 1:1 000 000 in spite of the foregoing discussion the scale error is so small that for all practical purposes the chart is considered to be a constant scale chart. The printed graduated scale line on one million training charts emphasises the point. This scale variation does become important as the chart scale becomes smaller. On a six million chart different graduated scale lines for different latitudes may be given and distance measurements must be carried out like Mercator.
Convergence. On Lambert's the value of convergence between two given meridians is given in the formula:

$$\text{convergence} = n \times \text{ch long}$$

where n is the sine of the parallel of origin. Parallel of origin, or n, is a factor in the orthomorphic formula upon which the mathematical adjustment is based. Parallel of

origin occurs in a position slightly displaced towards the pole from the parallel midway between two standard parallels. Its position is shown in fig. 6.4. Value of n,
however, is printed on the chart, but in the absence of this information, as pointed
out above, the sine of the mid latitude of the two standard parallels may be used for
calculating the convergence.

Example: Find convergence on a Lambert's projection between 119°W and 123°W at
43°N, given 55°N is the parallel of origin (fig. 6.4).

$$C = \text{ch long} \times n$$
$$= 4 \times 0.8192$$
$$= 3.2768°$$

Referring back to fig. 6.4, it will be noticed that the convergence between the
same two meridians, that is 119°W and 123°W at any other latitude, say, 40°N will
still be the same value of 3.2768°. In other words, on a Lambert's, convergence is
constant, its value depending on the value of n.

Next question is, is convergence correct? Well, if convergence is constant, it cannot
be correct, because the earth's convergency varies as sine of the latitude. In the above
example at parallel of origin (55°N) the earth convergency is

$$C = \text{ch long} \times \text{sin lat}$$
$$= 4° \times 0.8192$$
$$= 3.2768°$$

which is the same value as on Lambert's. The reason is, on a Lambert's the *convergence is shown correct only at the parallel of origin*. At 40°N the earth convergency:

$$C = \text{ch long} \times \sin 40°$$
$$= 4° \times 0.6428$$
$$= 2.5712°$$

whereas on Lambert's the convergence at 40°N is still 3.2768°, which is too high.
Similarly, towards the pole from the parallel of origin, the chart convergency is too
low.

To summarise, chart convergence is constant but not correct anywhere on the
projection except at the parallel of origin. Towards the equator the chart convergence
is too high; towards the poles it is too low.

Great circles. Say, a straight line is drawn on the earth, joining position A (55°N
123°W) to position B (55°N 119°W). The convergency between A and B on the
earth is 4 × 0.8192 = 3.3°. On a chart, if a straight line is to be a great circle, the two
positions must produce the same chart convergence as on the earth. Thus, a line joining A and B on the chart will be a great circle. Elsewhere on the chart, the convergence
between any two places will be different from that on the earth and such a straight
line will not be a great circle. A great circle, away from the parallel of origin appears
as a curve, concave to the parallel of origin. As the difference between earth convergency and chart convergency increases with distance from the parallel of origin,
so must the deviation between a straight line and great circle. However, for practical
purposes, a straight line drawn on a Lambert may be considered to be a great circle.
See fig. 6.5.

Further, since the convergence on the chart is different from that on the earth
(except at the parallel of origin) a radio wave travelling a great circle path will make

different angles on the earth and on the chart at a given meridian. Let us take an example. A radio wave leaves position 43°15′N 07°30′W on a bearing of 078° and crosses 10°E at 44°36′N.

The angle made by the radio wave at 10°E on the earth is

$$C = \text{ch long} \times \sin \text{mean lat}$$
$$= 17.5° \times \sin 4355\tfrac{1}{2}$$
$$= 17.5° \times 0.6936$$
$$= 12.14°$$

∴   Bearing on the earth $= 078° + 12.14°$
$$= 090.14°$$

On a Lambert with parallel of origin, say, at 58°N, the bearing:

$$\text{convergence} = \text{ch long} \times n$$
$$= 17.5° \times \sin 58$$
$$= 14.84°$$

and the bearing measured
$$= 078° + 14.84°$$
$$= 092.84°$$

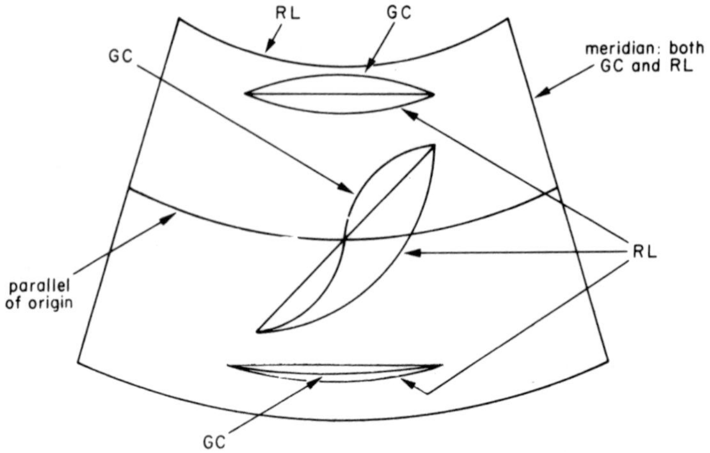

Fig. 6.5

Rhumb lines. Rhumb lines cut all meridians at the same angle. Therefore they will appear as curves concave to the pole, taking on the same curve direction as the parallels which are themselves rhumb lines by definition. Meridians are both GCs and RLs, so is the equator.

Fig. 6.5 shows the relationship between a straight line, rhumb line and a great circle.

Shapes. Provided the north–south coverage is not excessive and a careful choice of the standard parallels is made, shapes are sensibly preserved over the whole area.

Fits. Sheets will fit north/south and east/west if the scale and the standard parallels are the same. Also, a particular sheet of chart you are using need not have standard parallels on it. It may be a section of a larger projection.

Spacing of the standard parallels

If the scale of a projection is so adjusted that the maximum scale reduction on it equals approximately the maximum scale expansion, the overall scale error will be small. If the overall error does not exceed anywhere about 1% or so, the chart is considered to be a constant scale chart.

By placing the standard parallels about two-thirds of the total north–south coverage, that is, about 1/6th from the northern and the southern limits of the coverage, an even distribution of the scale error is achieved over the complete projection. The standard parallels on the Lambert's orthomorphic projection are placed according to this rule, see fig. 6.6.

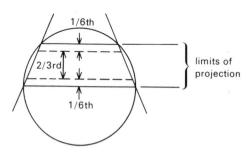

Fig. 6.6

Limitations of Lambert's projection

As explained earlier, the limitation of the projection lies in the north–south coverage. A good scale will be available and orthomorphism will remain intact as long as the ch lat between the two standard parallels is not too large. The greater the latitude difference between the two standard parallels the greater the scale error and distortion of the shapes. And if the shapes are greatly distorted, so are directions, with a loss of orthomorphism.

Uses of Lambert's projection

This projection is increasingly being used as a plotting chart. Its other uses are as topographical charts, radio aid charts (Consol, Decca, Loran), airways charts (Aerad) and meteorological synoptic charts.

Plotting on Lambert's projection

1. When measuring a track, use the mid-meridian between two points, and use this track angle to calculate the heading. This will be the mean great circle track, and also the rhumb line track as the measurement at the mid-point of a tangent to the curving RL track will almost be the same as the straight line GC track.
2. As always, use the latitude scale for measuring distance, *keeping the habit* of measuring in the area where the line is.
3. When finding W/V; the track and GS method is simpler. This method must be used when finding winds on a track which has a large E–W component (maximum

convergency is experienced). If the air plot method is used, measure the wind angle from the meridian nearest the fix.

4.   Bearings given by the station (QTE, QDM, QDR) and VOR bearings are measured at the station and plotted from the station meridian. Therefore, no correction for convergence is called for. Where the work is done at the aircraft (radio compass bearings), bearings are measured with reference to the meridian at the aircraft's position involving the aircraft's heading. Such bearings must be corrected for convergence *before* obtaining the reciprocal to plot from the station. Alternatively, the aircraft's DR meridian may be transferred to the station position, and the bearings plotted from the transferred meridian without correction for convergence.

Examples

Hdg	045		Hdg	045
Rel Brg	100		Rel Brg	100
GC Brg	145(T)		GC Brg	145(T)
Convergence	+ 2 (Towards eq)			−180
	147		Plot	325° from station position with
	−180			reference to the aircraft's
Plot	327° from station position			transferred meridian.

Practice problems

1.   Aircraft in DR position 43°N 81°W obtains QDM 173°, from station A (39°N 84°W) where variation is 36°E. What would you plot?
Answer: 029°.

2.   Aircraft on heading 270°(T) in DR position 46°N 84°W obtains relative bearing 239° from station 40°N 81°W. What would you plot?
Answer: 331°.

3.   Aircraft in DR position 69°30'S 65°W obtains QDM 277° from station at 71°S 70°W. Variation at a/c position 19°W; mean variation 22°W, variation at station 25°W. What would you plot?
Answer: 072°.

4.   Aircraft in DR position 72°S 66°W on heading 359°(T) obtains relative bearing 334°, Q.C. + 1, from station at 68°S 70°W. What would you plot?
Answer: 158°.

5.   Aircraft on heading 090°(T) in DR position 69°S 68°W
  (i) obtains relative bearing 090° from station at 72°S 68°W. What would you plot?
  (ii) obtains relative bearings 359° from station 69°S 64°W. What would you plot?
  (iii) obtains QDM 045° from station 69°S 64°W. Variation 20° at the station, 24° at the aircraft, mean variation 22°, all variations are easterly. What would you plot?

Answers: (i) 360°
       (ii) 265°
       (iii) 245°.

7: Polar Stereographic Projection

This is a perspective projection, with a light source at one pole and the plane of projection tangential to the other pole.

Since 360° of the Earth are represented on paper by 360°, the constant of the cone, n, is 1, if you care to consider it that way.

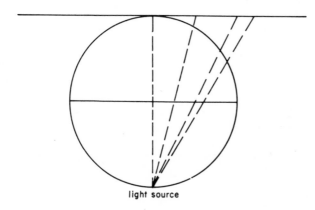

light source

Fig. 7.1

Construction
Select a convenient point to represent the pole, i.e. the point which is tangential to the model Earth in the basic concept of the projection.

From this point, draw in radial lines representing meridians at an angular spacing equal to the actual ch long.

With the Pole as centre, draw the circles of parallels. The radius of each parallel is calculated from the formula:

$$\text{Radius} = 2R \tan \tfrac{1}{2} \text{ co-lat}$$

where R is the radius of the model Earth; this model Earth radius in practice is derived from the Earth's actual radius of 250 000 000 inches divided by the scale required.

Appearance of the graticule
It will be noticed that a coverage of more than 90° of latitude can be achieved; see the graticule shown overleaf in fig. 7.2.

Properties
<u>Scale.</u> The scale is correct at the pole. The distances measured very close to the pole

are very nearly correct distances. Away from the pole the scale expands at the rate of $\sec^2(\chi/2)$, where χ is the co-latitude $(90° - \text{lat})$. Thus, although near the pole the scale may be considered to be constant, elsewhere distance measurements must be made with reference to the latitude scale as on a Mercator.

Although the scale varies from point to point, at any given point the scale is the same and this valuable asset will be recognised as one of the requirements of orthomorphism. We will now see how the scale varies north–south and east–west from a point.

First, the scale along a parallel will be considered. To get an idea of the scale expansion, we can simply compare the length of a parallel projected on the chart with the length of the same parallel on the earth.

The radius of a parallel on the polar stereographic is given as $2R \tan(\chi/2)$. Multiply this by 2π, that is, $2\pi \times 2R \tan(\chi/2)$, and this is its circumference, or the total length. The total length of a parallel on the earth is $2\pi R \cos \phi$, where ϕ is the given parallel. As $2\pi R$ occurs on both sides, we can compare $2 \tan(\chi/2)$ with $\cos \phi$.

lat, ϕ	χ	$\dfrac{\chi}{2}$	$\tan\dfrac{\chi}{2}$	Projection $2\tan\dfrac{\chi}{2}$	Earth $\cos\phi$
80°	10°	5°	0.087 49	0.174 98	0.173 65
70°	20°	10°	0.176 33	0.352 65	0.342 02
60°	30°	15°	0.26794	0.535 90	0.500 00

The ratio of the projected length to the length on the earth is the measure of the scale expansion or the exaggeration. Let us take 60° as an illustration.

$$\text{Exaggeration} = \frac{0.535\,90}{0.5} = 1.071\,80$$

(Note: the expansion we worked out just now is the same as that obtained directly from use of the scale factor mentioned above, that is, $\sec^2(\chi/2)$.) This shows that anywhere along 60° the scale expansion occurs at the rate 1.071 80.

We will now take any point on 60°N and calculate the scale from this point for an extremely small distance in the north–south direction. Extremely small though they must be, being limited to conventional mathematics, we will calculate the distance along the meridian from 60°02′N to 59°58′N.

The expansion along a meridian may be given by the relationship:

$$\frac{\text{length of the meridian between two parallels}}{\text{length of the meridian between the same parallels on earth}}.$$

We know that the length of a meridian on the earth is $\dfrac{\text{ch lat}}{360} \times 2\pi R$; the length on the projection is simply the difference between the radius of the two parallels.

$$\frac{\text{dist. on projection}}{\text{dist. on the earth}} = \frac{2R\left(\tan\dfrac{\chi'}{2} - \tan\dfrac{\chi''}{2}\right)}{2\pi R\dfrac{\phi' - \phi''}{360}}$$

$$= \frac{\tan\dfrac{\chi'}{2} - \tan\dfrac{\chi''}{2}}{\dfrac{\pi(\phi' - \phi'')}{360}}$$

We will work the numerator first and for sake of accuracy, use 9 figures.

$$\phi' = 60°02'; \quad \chi' = 29°58' \quad \frac{\chi'}{2} = 14°59'$$

$$\phi'' = 59°58'; \quad \chi'' = 30°02' \quad \frac{\chi''}{2} = 15°01'$$

Numerator

$$\tan 15°01' = 0.268\,260\,988$$
$$\tan 14°59' = 0.267\,637\,442$$
$$\text{subtract} \quad 0.000\,623\,546$$

Denominator

$$\phi' - \phi'' = 4' = 0.066\,66°$$

$$\frac{\pi \times 0.066\,66}{360} = 0.000\,581\,771$$

$$\frac{Numerator}{Denominator} = \frac{0.000\,623\,546}{0.000\,581\,771}$$

$$= 1.071\,80$$

This compares with the result we got earlier and clearly shows that the scale at 60°N is the same in all directions. If you now do the same exercise for another parallel, say, 60°02', you will work in entirely different sets of figures but the conclusion will be the same.

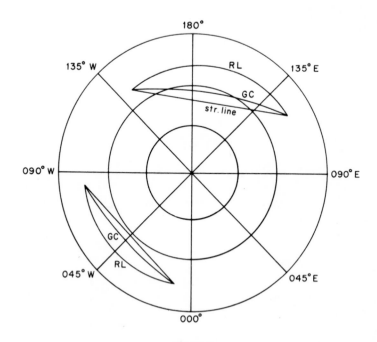

Fig. 7.2

Orthomorphism. Since scale varies in the same proportion just noted in all direction from a given point i.e. along the meridians and parallels, and since meridians and parallels cut each other at right angles, the chart is orthomorphic.

Convergency on the chart is the value of ch long, or 1 (one) degree of ch long is shown on the chart by 1° angular difference. On the Earth this difference exists only at the Poles, and only there on the chart is convergency correct. Elsewhere on the Earth, convergency reduces towards the Equator, where it is 0; so chart convergency is constant at 1, correct only at the Poles, too large away from the Pole.

Great Circle cannot be a straight line, since convergency is not correct. However, when plotting in high latitudes not too far away from the Pole where convergency is correct, a GC is taken to be a straight line for practical purposes. As distance from the Pole increases, a GC in fact curves concave to the Pole, and finally the Equator, itself a GC, will appear as a distinct curve. Rather an academic point for a chart used for aviation in the polar regions, perhaps.

Rhumb Line is a curve concave to the Pole. Parallels of latitude are demonstrative.

Equal Area only very nearly for small areas; from the properties of scale and convergency already mentioned, it is clear that the chart cannot be equal area.

Plotting radio bearings

Because a straight line approximates a GC, QTEs, QDMs, Consol bearings et alia may be plotted direct from the meridian of the station, after the usual resolution into the bearing True which the station measured.

Radio compass bearings, though, measured from the aircraft's meridian, must be resolved into an angle to be plotted from the station's meridian. Convergency must be applied before the reciprocal is found, its value being ch long.

Problems

Since convergency is 1, the sums are of no real profundity, only off-putting because the chart is not widely familiar.

1. What is the approximate GC bearing of B (71°N 50°E) which would be found by measuring the direction of A (71°N 90°E) on a straight line joining A to B? See fig. 7.3

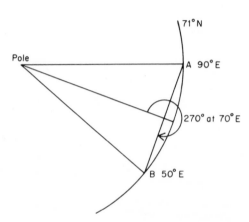

Fig. 7.3

The mean approximate GC bearing must be measured at the mid-meridian, as above. Both places are on the same parallel.

> Convergency between A and B is ch long $= 40°$
> ∴ convergency between mean meridian and A $= 20°$
> and GC bearing of B from A $= 270 + 20$ $= 290°$

2.   What is the highest latitude a straight line from A (70°N 00°) to B (70°N 80°W) will attain?

The distance, pole to A is 20 x 60 = 1 200 nm; call it PA.

The distance, pole to B is similarly 1 200 nm; call it PB.

Let C be the bisector of the line AB.

Join PC (pole to the point C).

Then distance PC represents the highest latitude — draw a little sketch.

∠PAC is 50° (angle BPA = ch long = 80°; ∴ ∠CPA is 40°, which makes ∠PAC = 50°) and knowing the distance PA,

$$PC = (1\ 200 \sin 50°)\ nm$$
$$= 919\ nm \quad \text{from the pole.}$$

Converting this into degrees and minutes, 919 nm is equivalent to 15°19′ and the latitude of C = 90° − 15°19′ = 74°41′N.

3.   What is the convergency between A (75°N 60°W) and B (75°N 10°W):
> a. on the Earth
> b. on the Polar stereographic?

Answer a. 48.29°
> b. 50°.

8: Transverse Mercator

This is a cylindrical projection, like a Mercator's but it differs from Mercator in that the cylinder is presumed to be tangential to any chosen meridian and its anti-meridian (fig. 8.1). The meridian to which the cylinder is tangential is called the Central Meridian (CM). The point where CM and the Equator intersect is called the point of origin.

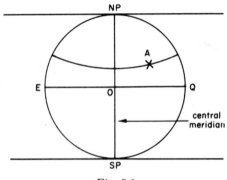

Fig. 8.1

Principle of construction
A cylinder is presumed to be wrapped round the reduced Earth (model globe reduced to the scale), being tangential to a chosen meridian with a light source at the centre of the Earth. The graticule resulting from such arrangement is modified mathematically to give orthomorphism. In practice, the chart is constructed mathematically throughout. A point to be projected on the chart is resolved in terms of N/S (N or S of the Equator) and E/W (E or W of the central meridian) co-ordinates and these co-ordinates (normally in inches) are then plotted on the graticule.

Suppose point A in fig. 8.1 is to be plotted on transverse mercator graticule (fig. 8.2). A GC is drawn up in the first instance from G passing through A, inter-

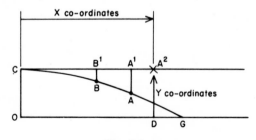

Fig. 8.2

secting CM at 90° at C. 0 is the point of origin and G is a point on the Equator, 90°
removed from 0. Now, if you think of the curve GAC in terms of a meridian on a
Mercator, with pole at G, you will appreciate that when the cylinder is unwrapped A
will appear at A^1. Similarly if there was another point B on the same GC it will appear
at B^1. If you are wondering why on an ordinary Mercator no such GC are plotted
initially, the reason is that point A on an ordinary Mercator would be on a meridian
which is already a GC by definition, whereas A on this projection is on a parallel of
latitude which only represents rhumb line distances.

The shifting of A on the Earth to position A^1 on the chart causes expansion of the
scale in the N/S direction (exactly as the meridians expanded in E/W direction on an
ordinary Mercator). This expansion must be compensated by extending the distance
CA^1 to CA^2 to obtain orthomorphism. This extension is calculated mathematically.
The co-ordinates to be plotted are: CA^2 (called X co-ordinates) and DA^2 (called Y
co-ordinates).

Graticule

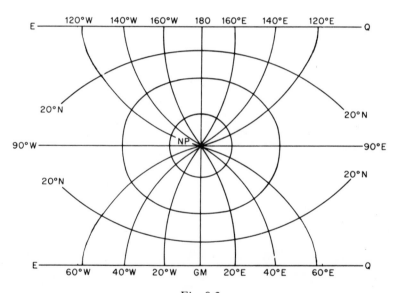

Fig. 8.3

The appearance of the graticule is shown in fig. 8.3. Note that the central meridian
appears as a straight line. It is a GC by definition. The two meridians 90° removed
from the CM also appear as straight lines (90° E and W in above figure). They are
GC by definition as well as by construction. (In fig. 8.2 notice that the curve of GC
is projected as a straight line perpendicular to the CM.) The Equator appears as a
straight line — in three places, in between the two Poles and at two extremities where
the projection provides full N/S coverage. Other parallels appear as ellipses, almost
circles near the Poles. Since on this projection both the meridians and the parallels
(with exceptions as shown above) appear as curves perhaps this feature clearly
distinguishes this projection from the ones we studied previously. This particular
feature, however, may not be discernible on small sheets.

Just as on a Mercator, on this projection a point 90° removed from the point of origin in an E/W direction on the Equator cannot be projected.

Properties

Orthomorphism. Yes, the meridians and the parallels cut each other at 90° and the scale variation from a given point is the same in E/W direction as in N/S direction by construction. This makes the chart orthomorphic.

Scale. The scale is correct along the central meridian. Expands away from the CM at the rate of secant of the great circle distance.

Convergency. Convergency is correct at the Equator and the Poles. Elsewhere it is incorrect. However up to 1 200 miles from the CM the error is very small and may be neglected.

Great Circles. Any straight line which cuts the central meridian at 90° is a great circle by construction. Elsewhere it is a complex curve as shown in fig. 8.4. For practical purposes any straight line drawn near the CM may be taken to represent a GC.

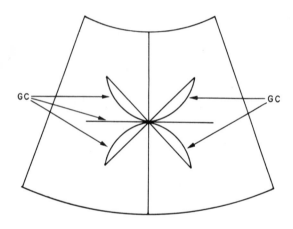

Fig. 8.4

Rhumb Line. All meridians and parallels are rhumb lines. Examine how they appear in fig. 8.3. Elsewhere a rhumb line is a curve concave to the nearer pole.

Shapes. These are correct for small areas.

Fits. You can attach another chart to your chart if the track goes beyond your chart coverage, but only in the N/S directions and provided both charts have the same central meridian and the same scale. There is a rolling fit in E/W directions for two charts of the same scale.

Uses. The projection is quite accurate up to about 300 nm of the CM. This makes it ideal for use as topographical maps of countries having large N/S extent (Italy, New Zealand, U.K.). The projection is not suitable for use as plotting chart due to curving meridians and parallels except in the polar regions where the meridians are still straight lines.

9: Grid Technique

In previous chapters we discussed Mercator, Lambert, Polar Stereographic and Transverse Mercator projections. Each one has certain advantages and disadvantages. Mercator is ideal for flying constant track angles but this involves flying RL distances. Further, radio bearings cannot be plotted directly. These two disadvantages practically disappear on Lambert and Polar Stereographic, but the pilot has to contend with changing tracks due to convergency of the meridians on the chart. Ideally we want a chart on which a straight line is a great circle, scale is considered constant and the meridians do not converge. No projection could provide all these properties together. However, by gridding a Lambert or a Polar Stereographic (and a Transverse Mercator in polar regions) we could combine all these on a single chart.

The first step towards gridding a chart is to select a suitable meridian as the *Datum*. The grid meridians are then drawn up parallel to and at a constant distance from the datum meridian. We now have two sets of directions on the chart: geographical meridians indicating true north and grid meridians indicating grid north (fig. 9.1).

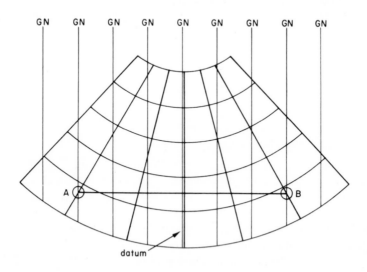

Fig. 9.1

Convergence of meridians
It will be noticed in fig. 9.1 that only one grid meridian (i.e. datum) coincides with true meridian; away from the datum there is an angular difference between the two,

the difference increasing with distance from the datum. This angular difference between True North and Grid North is called Convergence.

Relationship between TN and GN

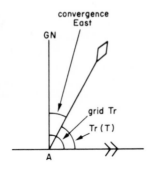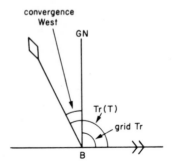

Fig. 9.2

Where the track starts at the datum meridian, the Grid track and True track will be the same initially. The grid track, however, will remain the same throughout the flight. Where a flight commences elsewhere e.g. from A to B in fig. 9.1, it is necessary to establish the relationship between grid and true tracks. Positions A and B are reproduced in fig. 9.2 above, showing the disposition of directions grid and true north. The grid track in both cases is 090° and if the convergence at A and B is, say, 5°, true track at A is 085° and at B 095°. This relationship may be expressed in the following formula, the convention being,

> Where true north lies to East of Grid North, Convergence is East
> Where true north lies to West of Grid North, Convergence is West

$$\text{Direction True} \quad \begin{array}{c} + \text{ Convergence E} \\ - \text{ Convergence W} \end{array} = \text{Direction Grid}$$

and, $$\text{Direction Grid} \quad \begin{array}{c} + \text{ Convergence W} \\ - \text{ Convergence E} \end{array} = \text{Direction True}$$

Example

Heading Grid is 150°, Convergence 10°W, What is Heading True?

Hdg(T) = Hdg(G) + Convergence W
= 150 + 10
= 160°

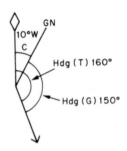

Fig. 9.3

Problems

1.   Hdg(T) is 258°, Convergence 20° E, What is Hdg(G)?
Answer: 278°

2. Hdg(G) is 040°, Hdg(T) is 065°, What is the convergence?
Answer: 25°W.

Grivation and Isogrivs

We established the relationship between GN and TN in the above article. However, in mid-latitudes, aircraft still steer by magnetic compass and therefore it is necessary to establish the relationship between GN and MN as well. The angle between TN and MN is variation. The angle between GN and MN is called Grid Variation, abbreviated to <u>Grivation.</u> Its value for any particular place is obtained from the lines of equal grivation, called <u>Isogrivs.</u> Isogrivs might be printed on a gridded chart or they may be added to such chart by plotting co-ordinates from simple calculation. As regards grivation, the convention is:

If MN lies to East of Grid North, Grivation is East.
If MN lies to West of Grid North, Grivation is West.

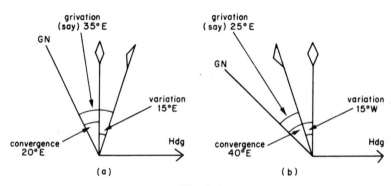

Fig. 9.4

From fig. 9.4 we establish the relationship in the following formula:

$$\text{Grid Direction} = \text{Magnetic Direction} \begin{array}{l} + \text{Grivation E} \\ - \text{Grivation W} \end{array}$$

Thus, by measuring the Grid heading and applying local Grivation, magnetic heading to steer is obtained.

<u>Grivation is the algebraic sum of convergence and variation</u>
In fig. 9.4(a),
    Grivation = 20°E + 15°E
           = 35°E
In fig. 9.4(b),
    Grivation = 40°E + 15°W
           = 25°E

When converting one heading to another, from above diagram, we establish the following rule:

$$\text{Hdg(M)} \begin{array}{l} + \text{E Var} \\ - \text{W Var} \end{array} = \text{Hdg(T)} \begin{array}{l} + \text{E Convergence} \\ - \text{W Convergence} \end{array} = \text{Hdg(G)}$$

In fig. 9.4(a), if the magnetic heading is 075°, the grid heading =
    075° + 15°E(Var) = 090° Hdg(T) + 20°E (Conv) = 110° Hdg(G)

If you are converting in the opposite direction, that is from Hdg(G) to Hdg(M), reverse the sign. E throughout is minus, W is plus. In fig. 9.4(b), given Hdg(G) = 130°, then Hdg(M) =

130 − 40 (E conv) = 090° Hdg(T) + 15° (W Var) = 105° Hdg(M)

Alternatively, Hdg(M) may be converted directly to Hdg(G) and vice versa by application of Grivation. In fig. 9.4(a),

Hdg(M) = 075° + 35(E Griv) = 110° Hdg(G)

In fig. 9.4(b)

Hdg(G) = 130° − 25(E Griv) = 105° Hdg(M)

Construction of grid on Lambert's

Two steps are involved in the construction of grid on a plain chart: construction of grid itself and plotting the isogrivs.

Construction of grid is quite simple. Simply choose a convenient meridian as the datum meridian and draw up lines parallel to the datum at a convenient distance (usually 60 or 100 nm). It should be appreciated that only those charts could be gridded which are considered constant scale charts. Grid is completed by drawing up another set of parallel lines, same distance apart but at 90° to datum meridian if necessary for plotting purposes. The chart will then look like the one in fig. 9.5.

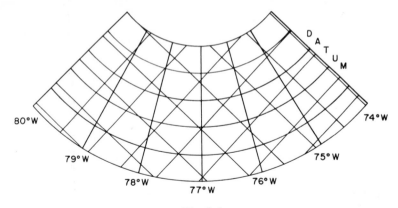

Fig. 9.5

Isogrivs contain two elements — variation and convergence. Variation lines (isogonals) are already on the chart. Therefore, in order to plot isogrivs, we will first need to draw up lines of convergence at 1° apart.

It will be remembered from the study of Lambert's that the convergence on this projection, (i.e. value of n) is always less than one. That is, one degree ch long on the Earth will appear at a distance less than one degree on the chart. For gridding purposes, we must calculate the distance from datum where one degree convergence will occur. This will invariably occur beyond one degree ch long printed on the chart. The calculation is simply carried out by dividing 1° (that is 60 minutes) by the value of chart convergence, that is, n.

Say, chart convergence was given as n = .7488. Then, 1° convergence on the chart =

$$\frac{60}{0.7488} = 1°20.1'$$

Plot the first line of convergence on the chart at distance of 1°20.1' from the datum. For example, if the datum was Greenwich meridian, one degree convergence occurs at 01°20.1'W; two degree convergence at 02°40.2'W and so on. These lines are lightly drawn on the chart so that they can be removed when isogrivs are drawn.

Next step is to calculate the value of grivation individually at every point where the line of convergence intersects isogonals on the chart. Finally, points having the same value of grivation are joined up by a smooth curve to give isogrivs. (Fig. 9.6.) The square grid is omitted for clarity.

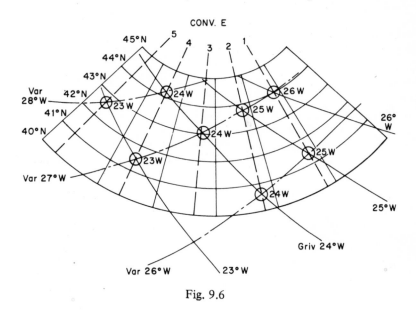

Fig. 9.6

Grid on polar charts

In polar regions, polar stereographic with a grid is generally used. A modified form of Lambert is also available and with its one standard parallel very near the pole it hardly differs from the other. Transverse Mercator may also be gridded in polar regions since meridians appear as straight lines radiating from the pole and curves of latitudes may be considered to be concentric. For all practical purposes near polar regions transverse Mercator may be considered to be a constant scale chart.

On these charts, convergence is 1; that is, one degree of ch long on the Earth appears at 1° distance on the chart. This makes construction of the grid easy in that lines of convergence need not be first constructed. Thus, if the datum is Greenwich meridian, then parallels of grid lines will cut 30W at an angle of 30°, 60W at an angle of 60° and so on (see fig. 9.7 overleaf).

On a complete polar grid, the direction from NP along the Greenwich anti-meridian is generally the direction of Grid North. Therefore, at the North pole, instead of having all true directions south only, we have a complete compass rose with reference to grid direction. Study fig. 9.8 to appreciate this.

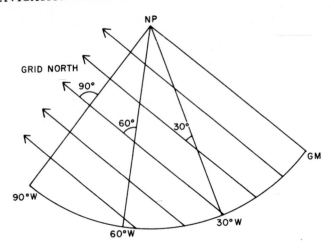

Fig. 9.7

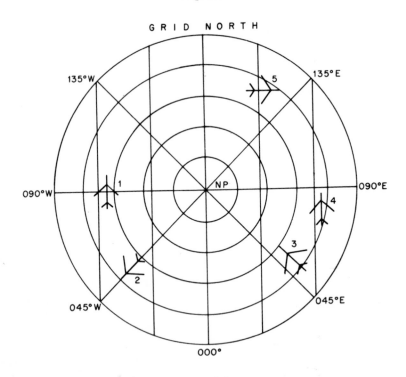

Fig. 9.8

In fig. 9.8,
Grid Heading of aircraft   1 is 000
                            2 is 225
                            3 is 315
                            4 is 000
                            5 is 090

As for the conversion of Grid value to True and reverse, since the value of n is 1 the conversion formula becomes

$$\text{Grid direction} = \text{True direction} \begin{array}{l} - \text{ Long E} \\ + \text{ Long W} \end{array}$$

$$\text{and}\quad \text{True direction} = \text{Grid direction} \begin{array}{l} + \text{ Long E} \\ - \text{ Long W} \end{array}$$

For illustration, the Grid Heading of aircraft 2 in fig. 9.8 is 225°

Its true heading   = 225 − 45 (Long W)
                                 = 180°

True heading of aircraft 3 (heading along meridian to NP) is 360°. Therefore its Grid Heading   = 360 − 45 (Long E)
                                 = 315°

These signs must be reversed in a South Polar Grid − See typical problem 3 below.

In the lower regions where the magnetic compass may still be used, isogrivs must be first constructed. These are again simpler since in this case the value of grivation is the value of ch long in whole degrees combined with the value of local variation.

Transverse Mercator outside polar regions

The task of gridding in such areas becomes difficult due to curving meridians. If it is necessary to grid such chart, convergence should be determined at numerous points all over the projection, before following up the normal steps.

Typical grid problems

1. An aircraft in position 40°N 10°E has a magnetic heading of 150° and a grid heading of 170°. If variation is 10°W and n = .8, what is the datum meridian?

a.   Hdg(T) = 150 − 10°W (Variation)
           = 140°

b.   Convergence = Hdg(G) − Hdg(T)
                = 170 − 140
                = 30°E (See fig. 9)

c.   Ch long from Datum = $\dfrac{\text{Convergence}}{n}$

            $= \dfrac{30}{.8} = 37\tfrac{1}{2}°$

d.   Therefore Datum 10E + 37½E
                = 47½°E

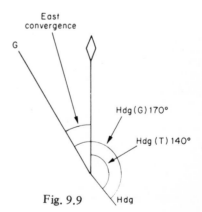

Fig. 9.9

2. Using Grid based on 20°W, what will be the magnetic heading of an aircraft in position 50°E, given variation 8W and n = .75. Grid heading of the aircraft is 224°.

a.   Convergence = n × ch long
                = .75 × 70
                = 52½°W (see sketch)

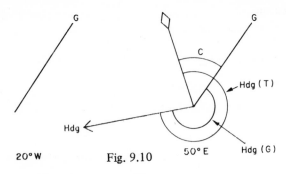

Fig. 9.10

20°W   50°E   Hdg (G)

b. Hdg(T)   = Hdg(G) + Convergence W
             = $224° + 52\frac{1}{2}°$
             = $276\frac{1}{2}°$
c. Hdg(M)   = $276\frac{1}{2}° + 8°$ W (Variation)
             = $284\frac{1}{2}°$

3.   An aircraft on south polar grid in position 75°S 20°W has a grid heading of 210°.
What is its true heading?

Hdg(T) = Hdg(G) + Long W
       = 210 + 20
       = 230° (see fig. 9.11)

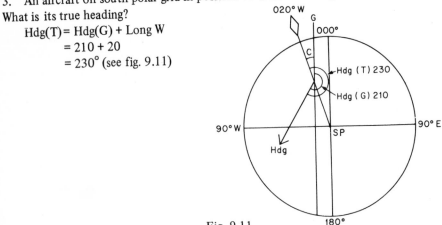

Fig. 9.11

4.   An aircraft on North polar grid is steering a true heading of 080°(T) and a grid
heading of 140(G). What is its longitude?

$$Hdg(G) = Hdg(T) \begin{matrix} - E \text{ Long} \\ + W \text{ Long} \end{matrix}$$

$$\therefore Hdg(G) - Hdg(T) = \begin{matrix} - \text{Long E} \\ + \text{Long W} \end{matrix}$$

140 − 80
= + 060
= 060W (see fig. 9.12)

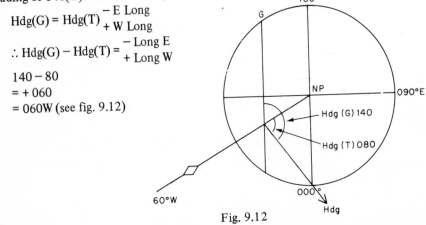

Fig. 9.12

Practice problems

1. Hdg(T) is 258°, Convergence 20°E; what is Hdg(G)?
Answer: 278°.

2. Hdg(G) is 040°, Hdg(T) 065°; what is the Convergence?
Answer: 25°W.

3. An aircraft on North Polar grid is steering a true heading of 080° and a grid heading of 140°. What is its longitude?
Answer: 60°W.

4. An aircraft has a grid heading of 310° using a chart based on grid datum of 40°W. If the variation is 10°E, n = .8 and Hdg(M) 340°, what is the aircraft's longitude?
Answer: 10°E.

5. Grid datum meridian is 50°W, n = .7, aircraft's position 50°N 20°W. If the grid heading is 257° what is the magnetic heading, given the variation is 8°W?
Answer: 286°.

AIRCRAFT INSTRUMENTS

10: Airspeed Indicator (ASI)

The principle of an Air Speed Indicator (ASI) is the measurement of two pressures called Static and Pitot pressures.

If you move an open ended tube through the air, pressure will be exerted at the closed end of the tube. This pressure is composed of two components: that pressure which would be present at the closed end of the tube irrespective of whether the tube is stationary or moving. This is due to the atmospheric pressure and is called 'static' pressure. The other component is the additional pressure entirely due to the tube's movement through the air. The faster the tube is moved, the greater the pressure that is exerted. This pressure is known as 'dynamic' pressure. The sum total of the two components is known as 'pitot' pressure. That 'additional' or dynamic pressure mentioned above is the one representative of the air speed of the tube and the one we are interested in. But the pressure produced at the closed end of the tube is pitot pressure. Thus, to have an indication of the air speed it is merely a question of extracting dynamic from pitot pressure.

Pitot pressure, P, equals dynamic pressure, D, plus static pressure, S:

or, $P = D + S$

$\therefore \underline{D = P - S}$

An ASI measures the dynamic pressure by solving the above formula, that is, by subtracting static pressure from pitot pressure continuously through the flight and presenting the information in terms of the aircraft's air speed.

Construction

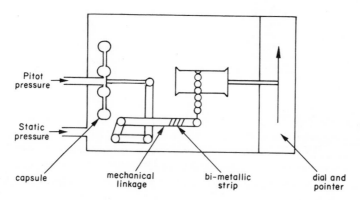

Fig. 10.1

The construction of an ASI is shown in fig. 10.1 above. In place of an open ended tube, the ASI uses an open ended capsule, fixed inside an airtight case. The open end of the capsule is connected to a tube installed in the aircraft's nose or on one of its wings, and facing the air flow. This tube is called a pitot tube or sometimes, pressure head. Thus, the pressure produced inside the capsule is the pitot pressure and the other face of the capsule will expand or contract in response to the variations in this pressure. This pitot pressure is in fact passed through a short capillary before feeding into the capsule: this is to ensure a smooth supply of pressure, damping out any surges. Inside the case (but not inside the capsule) static pressure is fed. The result of the presence of this static surrounding the capsule is to check the expansion of the capsule since the face of the capsule expanding under pitot pressure must overcome the opposition due to this static. The resultant expansion will, therefore, be equal to the value of P − S which is dynamic pressure.

All that remains to be done now is to transmit this capsule face movement (expansion and contraction) to a pointer. This is done by means of a suitable mechanical linkage. We are not interested in the details of the interior mechanism except that somewhere in the mechanical linkage a bi-metallic strip is introduced to compensate for expansion/contraction of the linkage due to temperature variations.

Calibration

This is a problem, where, having constructed an ASI one asks oneself, 'Where, on the face of the dial, am I going to mark off 60 kt and 100 kt and the rest of the speed range?' The problem is not a simple one since here we are dealing with atmospheric pressures. You may, for example, take the ASI for calibration in a car. Run the car up to 60 kt, and where the pointer reaches at that speed, mark off 60 kt. But the question is, will the pointer reach the same position when the car is doing 60 kt the following day? Or, for that matter, the following hour? The answer would be 'no', for the pressures vary from hour to hour and day to day. It is, therefore, necessary first to adopt some standard in calibrating the ASI and then make allowance for known departure from that standard. The standard adopted is the International Standard Atmosphere (ISA) air density at sea level. This is that density which is produced when the pressure is 1013.2 mb and the temperature is +15°C. The calibration formula is:

$$P = \tfrac{1}{2}\rho \, V^2 \left(1 + \frac{V^2}{4C^2}\right)$$

where   P = pitot pressure
ρ = air density under ISA
V = indicated air speed
C = speed of sound

Errors

An ASI suffers from the following four errors:

1.   Instrument Error. This is due to small manufacturing imperfections and the fact that a very minute movement of the capsule is expanded to give a reasonable movement of the pointer. The extent of this error can be determined by comparing the readings at various air speeds against a standard ASI. The errors so found are recorded on a card. The reading you take from the Air Speed Indicator is known as 'Indicated Air Speed' (IAS).

2.   Pressure Error. This error is also known as 'Position Error'. The pressures presented inside the capsule and inside the case must be correct pressures if correct speed is to be measured. As far as the pitot tube is concerned to a large extent it gives

correct values. Static is the main source of the error since to be correct it must be entirely free from any air flow under pressure or from other disturbances caused by aerodynamic surfaces. If the static tube is combined with the pitot head, an arrangement shown in fig. 10.2, it will be seen that the static tube will be surrounded by

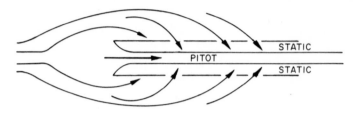

Fig. 10.2

disturbed air flow and some dynamic pressure will enter it giving erroneous results. These errors can be minimised if the static source is completely divorced from the pitot head. Under this arrangement it takes the form of a 'static vent' (as against static head). It consists of a small copper plate with a hole drilled radially through the centre. The air under static pressure enters a tube through this hole and is conveyed to the inside of the instrument case at the other end of the tube. An ideal position is found for installation of the static vent where the disturbance is minimum and to improve the results further the plate is fixed flush with the fuselage skin to avoid encouraging local disturbances. Approximately 95% of the pressure error is reduced by using this arrangement. Some aircraft use two static vents, one on each side: this is to balance out any errors in static pressure caused by yawing or side slipping.

There are two distinct types of pressure errors.

Position Error. This is the error entirely due to the location of the pitot head and static vent, and the magnitude of the error depends on this. As we saw above the static is the main source of the error and no matter how ideal a position for static vent is found there is bound to be some disturbance present around it. The magnitude of the error also depends on the aircraft speed, since, generally, the higher the speed the greater the disturbance. This error, like instrument error, is a known factor and can be determined throughout the ASI's speed range. In practice both instrument and position errors are combined together on a single card, called 'Pressure Error' (PE) Card. The reading taken directly from the instrument is called Indicated Air Speed (IAS). PE is applied to the IAS to give Rectified Air Speed (RAS).

Manoeuvre-Induced Error. This error occurs when the pitot head is not directly facing the air flow, for example, when an aircraft is in a climb or descent. The error also occurs when extraneous disturbance is caused by the pilot's actions such as lowering the undercarriage or manipulating flaps. The aircraft is then no longer aerodynamically 'clean' and the static will be affected. The situation is aggravated by the fact that short term pressure fluctuations are present during rapidly changing manoeuvres and that some time, however small, must elapse between arrival of the pressure at one end of the tube and its delivery at the instrument end of the tube. Thus you may have an indication which was true a few moments ago. These errors are random and therefore cannot be pre-determined for recording on a card.

To summarise, the magnitude of pressure errors depends on:

(a) the position of the static vent;

(b) aircraft speed;

(c) angle of attack and type of manoeuvre;

(d) aerodynamic state, e.g. whether flaps down etc.

3.   Density Error. This error occurs due to the calibration of the ASI. An Air Speed Indicator reads correct air speed only when flying in an air mass of such density as would be produced if the prevailing pressure was 1013.2 mb and the temperature was +15°C. These data describe the ISA density at the sea level but the conditions may prevail above the surface in its near vicinity. By 'near vicinity' we mean a few hundred feet rather than a few thousand feet. At 5 000 ft for example for this density to exist the temperature required is approximately −32°C and at 10 000 ft it is −73°C. Thus, it is most unlikely that an ASI would read correct air speed at 10 000 ft.

As for the nature of the error, as the height is gained the atmosphere becomes rarer and density decreases. Therefore, the dynamic pressure falls and the indicator gives a reading which is too low for a given air speed. The rectified air speed must therefore be corrected for density error to give the true indication of the speed and you will have noticed from above that in the majority of the cases it will be an additive correction. The correction is made on the computer by setting pressure altitude against temperature in the 'air speed' window and reading off True Air Speed (TAS) on the outer scale against rectified air speed on the inside scale. To summarise:

    IAS ± PE   = RAS;

    RAS:Density Error = TAS

TAS is the air speed the aircraft is actually doing through the air.

From the foregoing it will be appreciated that an aircraft flying straight and level from warmer air mass into colder air mass or vice versa will experience changes in the air speed indications. For example: an aircraft is flying at 10 000 ft and the temperature is −5°C, its RAS is 130 kt. The TAS for this pressure altitude, temperature and RAS is 150 kt. Later the same aircraft enters an air mass having temperature of, say, −15°C. This air mass is relatively colder and therefore denser, so the new dynamic pressure will be higher and the indications will change. On the computer, to maintain a TAS of 150 kt as before but under the new set of conditions, the RAS will show an increased reading of 133 kt − a higher indication. Alternatively if you are maintaining an RAS of 130 throughout the new TAS is reduced to 147 kt.

In absence of a computer, density error may be estimated from the following formula:

    TAS = RAS + (1.75% of RAS per 1 000 ft of altitude)

For example, RAS is 130 and height 10 000 ft:

    TAS = 130 + 1.75% of 130 per 1 000 ft × 10

        = 130 + 17.5% of 130

        = 130 + 22

    TAS = 152 which compares with above illustration.

4.   Compressibility Error. This error generally applies to high speed aircraft doing a TAS of 300 kt or over. At such speeds air compresses when brought to rest in front of the pitot head and consequently enters the tube under higher pressure, giving an overread of IAS. Thus, compressibility error is corrected as a subtractive factor. Air nearer the Earth's surface is not easily compressed since it is already dense and the

compressibility error near surface levels is negligible. The error increases with altitude. For a given altitude an increase in air speed increases the error. That portion of the calibration formula given in the brackets is known as the Compressibility Factor. It confirms the above two statements. The correction for this error is made on the computer. Frequently graphs for a particular aircraft are available which correct for this as well as pressure error in one operation. RAS corrected for compressibility = equivalent air speed (EAS). EAS corrected for density = TAS.

Blockages

Blocked Pitot. If the pitot tube gets blocked by ice or other obstruction before take-off there will be no reaction at all and the instrument will read zero. In level flight should the pitot get blocked the dial will hold the reading unless pressure leaks away in which case the pointer will return to zero. During a descent a blocked pitot will give an underread as the increasing static will put up increasing opposition to constant value pitot pressure trapped inside the tube.

If the pitot tube develops a leak the ASI underreads.

Blocked Static. Should the static become blocked during take-off the instrument will underread (high value static is trapped in). During level flight the instrument continues to give the same reading. During a descent it will overread and a dangerous situation can readily be envisaged when the aircraft could stall at an air speed indicated as being well above the stalling speed.

Serviceability checks

As far as the instrument itself is concerned there is nothing a pilot can check except that the pointer is not stuck on the dial. A pilot must check that the pitot head cover and the static vent pins are removed, that the pitot head is not bent, cracked, misaligned with the airflow or otherwise damaged in any way.

11: Altimeters

Pressure altimeter

The atmosphere has weight and this weight exerts pressure. An ordinary household
barometer measures this pressure and indicates the weather. If we took this barometer
to the top of a tall building it would be noticed that for the same weather conditions
the pressure indication is less. This is because the atmosphere remaining above the
barometer at height is less than at ground level. If we knew the rate of the fall of
pressure with height, we could graduate that barometer to read in terms of height
instead of reading the pressure. An aircraft altimeter is simply an aneroid barometer
adapted for use in aircraft on the basis that 1 millibar of pressure change takes place
for a change of height of approximately 27 feet. In the above illustration, if the
barometer reads 1010 mb at ground level and 1008 mb at height, the top of the
building is approximately 60 feet (30 ft to a millibar is a good figure to use unless
the examination question gives a specific figure).

Construction and operation

Two or three capsules each having vacuum or partial vacuum in them are used to
achieve sensitivity. They are stacked together with one face fastened down to the
base permitting movement due to pressure changes at the other end. (Fig. 11.1.)

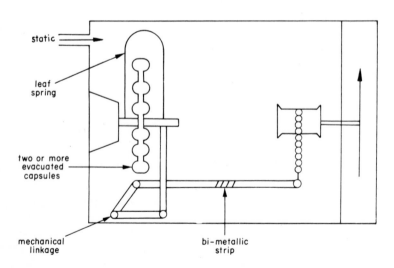

Fig. 11.1

To prevent the capsules from collapsing (because of the vacuum in them) a spring called the leaf spring is used to hold up the moveable face. The movement of the capsules in response to change in height is transmitted to the pointers, usually three pointers, through a suitable mechanical linkage. Somewhere in the linkage a bimetallic bar or other similar device is inserted to compensate for temperature variations. The whole assembly is encased in a container, having an inlet for the static pressure but otherwise airtight.

As the aircraft gains height the value of static decreases and the capsule expands under the tension of the spring. (Fig. 11.2.)

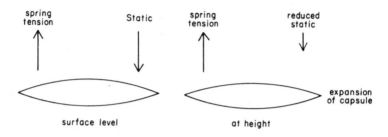

Fig. 11.2

Similarly, as the aircraft descends, static increases and the capsule contracts. These movements are magnified sufficiently to give reasonable pointer displacement, which move around a dial calibrated in feet.

Calibration

Like the ASI we are here dealing with constantly varying atmospheric pressures which show up so vividly on a household barometer. Therefore, a standard for calibration must be adopted and this standard is the International Standard Atmosphere as follows:

1.   The sea level pressure is presumed to be 1013.2 mb.
2.   Sea level temperature +15°C.
3.   The temperature falls with height at the rate of 1.98°C per thousand feet up to a height of 36 090 ft. Above this height the temperature remains almost constant at −56.5°C. Note that this item is not included in calibration of an ASI, and that this is not the complete definition of International Standard Atmosphere.

Errors

The instrument suffers from five different errors as follows.

1.   Instrument Error. This error is similar to that in an ASI except that one more factor must be considered. That is the presumption that the rate of fall of pressure with change in height is a constant value. In actual circumstances, as the atmosphere becomes rarer that rate of fall must decrease with height. To this extent the altimeter becomes unreliable when flying at high altitudes.

2.   Pressure Error. This error occurs for reasons similar to an ASI and both its components, position and manoeuvre induced errors, are present. As regards the error

due to the position of the static vent a P.E. card is available for correction to the readings taken from the instrument.

3.   Barometric Error. A household barometer again comes in handy to show that if it was calibrated to give 0 ft reading where the pointer stood one day, the next day the pointer would suggest that the house had gained or lost height. As for altimeters, any variations from the standard calibration conditions will give an error. This is known as barometric error, known in the early days as Pressure Error. It is most important that the altimeter hands are set to read height above the ground or mean sea level before taking off, thus correcting for pressure difference from 1013.2 mb. This is done by setting the appropriate pressure on what is called the 'millibar sub-scale', the sub-scale being visible through a window on the instrument. A knob is provided on the instrument to set the desired pressure. When it is turned the whole mechanism inside the instrument turns, turning the pointers with it. Various pressure settings available are considered later in the chapter. When correct pressure is set the indication is correct (except for inaccuracies due to other errors) and the indication remains correct as long as the surface pressure remains unchanged. When that pressure changes, the indication is in error again: this topic is dealt with in detail later in this chapter (Pressure—altimeter relationship).

4.   Temperature Error. As mentioned earlier the instrument presumes the surface temperature to be +15°C, reducing at the rate of 1.98°C per thousand feet. This lapse rate prescribes an exact temperature for every flight level from the surface (SL) to the aircraft level. Therefore, if the flight level temperature is not the ISA temperature, and the temperatures in the column of air below the aircraft are not the ISA temperatures, an error in the indication will occur.

The magnitude of this error is relatively small when at lower levels but at higher levels it does become significant. The corrections are carried out on the slide rule side of the navigation computer. Set pressure against the corrected outside air temperatures (COAT) in the window marked 'altitude'. Then, against indicated altitude on the inside scale, read off the true altitude on the outside scale. For example, when at FL 5 000 ft with COAT −12°C, the true altitude is 4 700 ft. It must be pointed out that strictly speaking, not the outside temperatures but the mean temperature of the air column below the aircraft must be used to obtain correction. In the absence of this information the outside temperatures are used to give us an approximate value.

5.   Time Lag Error. This error occurs due to the time that the pressures take, however small, to travel from one end of the tube to the other. (Pressures travel at approximately 1 100 ft/sec and on a modern aircraft the distance to travel may quite easily be of the order of 50 ft.) This error is most noticeable during steep climbs or descents. During a climb, higher pressure is present at any given instant and the altimeter underreads. Similarly during a descent, time lag causes an altimeter to over-read. Masked time lag errors are also present during other manoeuvres.

Blockage
Should the static vent become blocked through ice or other obstruction old static will remain trapped and height changes will not be indicated.

Pressure settings
Due to the nature of calibration and various other factors an aircraft altimeter rarely

reads correct height. With this limitation in mind we are ready to familiarise ourselves with various settings available. At the outset it must be emphasised that when a specific pressure is set on the sub-scale that pressure becomes the datum from which the difference in pressures between it and the actual pressure experienced at aircraft level is measured. It is this difference which positions the pointers.

Let us take an illustration. The datum set is 1 000 mb and, say, the pressure upon the aircraft altimeter is 700 mb. The difference of 300 mb between the datum and actual pressures resolves into a pointer displacement of 300 x 30 = approximately 9 000 ft. If the datum set was 1020 in the same circumstances the difference would be 320 mb and the pointer displacement of 320 x 30 = approx. 9 600 ft. From the illustration we conclude that if pressure setting is increased, the height reading is increased, and vice versa.

Various settings available are given names in Q code. These are:

QFE: the barometric pressure at the level of the aerodrome. When set the altimeter indicates height of the aircraft above the highest point on the manoeuvring area on the aerodrome. This setting is generally used for take off, landing and particularly when carrying out radar approaches. When set, the correction for the barometric and temperature errors is duly made and if the aircraft is stationary on the ground, only the instrument error is present.

This setting is also used for checking the altimeter serviceability.

QFF: the barometric pressure at aerodrome level, reduced to mean sea level. This reduction is made using the values of actual pressure and temperature conditions prevailing at the time and not the ISA values. This setting is mostly used by meteorological offices for plotting synoptic charts.

QNH: the barometric pressure at aerodrome level, reduced to mean sea level using ISA formula. When set, an altimeter calibrated in ISA reads height of the altimeter above mean sea level. This height is called Altitude. Again, when set, it is corrected for the barometric and temperature errors, and when on the ground it will correctly indicate the altimeter's height above MSL. In flight, however, even if the pressure remains unchanged, a temperature deviation from ISA is likely to be present, introducing temperature error.

This setting is also used for checking altimeter serviceability. Two types of QNH values are available — spot QNH and Regional QNH. Spot QNH is only valid for the spot (aerodrome) where the pressure reading took place. This may be used when taking off or landing as an alternative to QFE. The law requires a pilot taking off in controlled airspace to have at least one altimeter set to aerodrome QNH value.

Regional pressure setting (i.e. QNH) in contrast to spot or aerodrome QNH is applicable throughout the region for which it is given. For the purpose of providing this service the U.K. is divided into 14 regions known as Altimeter Setting Regions (ASRs). Each region produces, on the hour, a forecast of the lowest values in the area, and such forecast QNH is valid for one hour. Being the lowest forecast value, the setting is used for the purpose of maintaining adequate terrain clearance. When a lower value is set a lower indication results, and maximum safety is assured. The indication may not be a correct indication of the altitude but the error is in favour of safety. We said it may not be correct, because, first of all, it is a forecast value and secondly, the value chosen is true for only a small area in the region where lowest values were recorded. Elsewhere in the region, higher pressures may be expected.

Other reasons for an incorrect indication are instrument error, temperature error, and the pressure given being rounded off to a whole figure.

Standard Setting: strictly, this is the altimeter reading on the ground with 1013.2 mb set on the sub-scale, but is commonly taken to be 1013.2 mb setting, used for flight separation between aircraft flying under IFR. With this setting the altitude indicated is called pressure altitude and the level at which an aircraft flies is called flight level. For example, if 10 000 feet of pressure altitude is indicated, the aircraft's flight level is 100. It will be appreciated that the aircraft's altitude and pressure altitude will be the same only when the QNH value is also 1013.2 mb. At all other times the indication will be in error on a 1013.2 mb setting. However as between two aircraft flying close to each other and therefore in the same air mass, both will have similar error in indication. The separation can therefore be maintained accurately.*

Pressure–altimeter relationship

From the above it will be apparent that once an aircraft departs from a place with correct setting, the indication will remain correct (disregarding limitations of QNH mentioned above) only as long as the datum pressure continues to prevail. When new pressures are encountered, the indication becomes erroneous. For the sense of the error, the rule is: when flying from an area of high pressure to an area of low pressure the altimeter overreads. The rule may more easily be remembered from the phrase – 'high – low – high'. The reverse is also true. This is illustrated in fig. 11.3. The pressure at position A is 1000 mb:

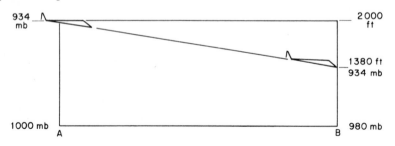

Fig. 11.3

and an aircraft is flying with this setting, its indication being 2 000 ft. The pressure upon the altimeter

$$= \frac{2\,000}{30} = 66 \text{ mb less than } 1\,000$$

$$= 934 \text{ mb}$$

In order to maintain an indication of 2 000 ft the aircraft must fly in the airspace where the pressure on the altimeter is 934 mb. Let us assume that the pressure at B is 980 mb. At B, the height at which pressure of 934 will occur = 980 − 934

$$= 46 \text{ mb}$$
$$= 1\,380 \text{ ft}$$

*For fuller discussion on this topic the student is advised to consult Ch. 3 of the companion volume 'Aviation Law for Pilots'.

Thus, the aircraft will actually be at 1 380 ft although the indication is 2 000 ft. The aircraft flew from an area of high pressure to an area of low pressure and high − low − high rule applied.

Temperature−altimeter relationship

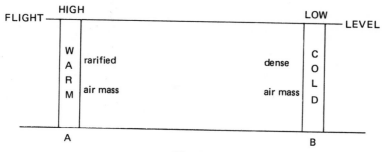

Fig. 11.4

In fig. 11.4 the area around A is warmer than the area around B. Therefore, at given level, the altimeter will experience higher pressure above it when over A than over B. Thus, flying from A to B without altering the datum will amount to flying from high pressure to low pressure: high−low−high: and the altimeter will overread. The reverse is also true.

Drift−altimeter relationship
From fig. 11.5 it will be observed that if an aircraft (aircraft A in figure) experiences persistent starboard drift in the northern hemisphere, that aircraft is approaching a low pressure area and the pilot must anticipate that his altimeter will read too high. Aircraft B with port drift is approaching an area of high pressure and the altimeter will underread. Reverse results are obtained in southern hemisphere.

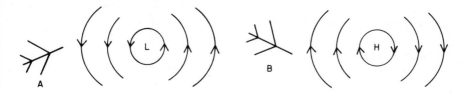

Fig. 11.5

Density altitude
This is defined as the height in ISA at which prevailing density will occur. Density altitude is mainly used in ascertaining the performance data of aircraft. It is the same as pressure altitude when local temperature is ISA temperature. If the local temperature is lower than ISA the density altitude is lower than the pressure altitude, and vice versa. 1 °C difference of temperature causes density altitude to separate from pressure altitude by approximately 119 ft.

Indicator

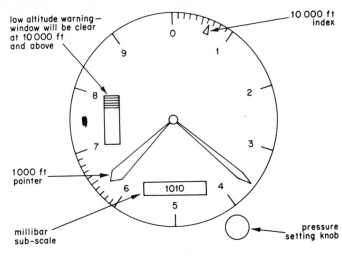

Fig. 11.6

A few sample problems using the various settings and their implications may be useful. Use 1013.2 mb as the standard setting, and 1 mb = 30 feet. And remember that if you reduce the value of the sub-scale setting, the height indication decreases; and conversely, of course, an increase in the setting increases the height indication. A quick 'situation sketch' is often very useful indeed.

1.   When flying at FL 55, will there be a minimum of 1 500 feet clearance over high ground, 3 850 ft amsl when the QNH is 1007.2 mb?

Answer: No.

Solution

QNE − QNH difference = 6 mb = 180 feet. When at FL 55, the true altitude is 5 500 ft − 180 ft = 5 320 feet. In order to maintain a 1 500 feet clearance your true altitude should be at least 3 850 feet + 1 500 feet = 5 350 feet.

2.   An aerodrome, 990 feet above mean sea level has QFE of 990 mb.

   (a) What is the QNH?

   (b) What clearance has an aircraft from a spot height in the vicinity of the aerodrome 3 000 ft amsl, if the aircraft is at FL 40?

Answer: (a) 1023 mb; (b) 1 294 ft.

Solution

(a) The altimeter reads 0 feet with 990 mb set. For QNH, wind it up until it reads the aerodrome's elevation that is, 990 feet. Thus, the number of millibars to add to 990 mb is

$$\frac{990}{30} = 33 \text{ mb}$$

and   QNH = 990 + 33 = 1023 mb.

(b) QNH − QNE difference = (1023 − 1013.2) mb
                                = 9.8 mb = 294 ft.

When 1013.2 set, the indication = 4 000 ft
QNH − QNE difference   =   294 ft
Indication when QNH set   = 4 294 ft
Elevation of the spot height   = 3 000 ft
Clearance   = 1 294 ft

3.   The en route spot height is 1 600 metres amsl, and the regional QNH is 995 mb. Assuming 1 millibar = 29 feet, what is the actual clearance over this spot height when flying at FL 70?
Answer: 1 224 ft.
Solution
   (i) 1 600 metres = 5 248 ft.
  (ii) QNE − QNH difference = (1013.2 − 995) mb
                            = 18.2 mb = 528 ft.
 (iii) Indication when QNE set = 7 000 ft
       QNE − QNH difference   =   528 ft
       Indication when QNH set = 6 472 ft
       Elev. of the spot height = 5 248 ft
       Clearance   = 1 224 ft

4.   An aircraft is flying at flight level 65. There is some high ground en route, the elevation of the highest point is 1 350 metres. What actual terrain clearance will this aircraft have when passing over this high ground if the QNH is 1003.2 mb?
Answer: 1 772 feet.
Solution
   (i) 1 350 metres = 1 350 × 3.28 ft
                = 4 428 ft.
  (ii) Difference between QNE and QNH = (1013.2 − 1003.2) mb
                              = 10 mb = 300 ft.
 (iii) Indication when 1013.2 set = 6 500 ft
                        =   300 ft*
       Indication when QNH set = 6 200 ft
       Elevation of the high grd. = 4 428 ft
       Clearance   = 1 772 ft

5.   An aircraft on a track of 089°(M) wishes to fly at the lowest flight level which will give it a clearance of 5 000 feet amsl. If the regional QNH is 992 mb, calculate the flight level.
Answer: FL 70.
Solution
   (i) QNH − QNE difference = (1013.2 − 992) mb = 21 mb.
                            = 630 ft.
  (ii) Indication on QNH   = 5 000 ft
       QNH − QNE difference =   630 ft
       Indication on QNE   = 5 630 ft

* We subtracted 300 feet because the QNH is lower than the QNE and therefore, the QNH indication is lower than 6 500 feet.

(iii) Therefore, the lowest available flight level is FL 60 and the flight level appropriate for the track of 089°(M) is FL 70 (odd thousands).

Servo altimeters

These are second generation altimeters designed to overcome generally some of the serious limitations of the conventional altimeters and to improve overall performance.

Capsules are still retained – generally two of them and evacuated of air – which expand and contract with changes in pressure. The rest of the mechanical linkage is replaced by a servo-assisted transmission system. The altimeters operate as follows.

The mechanism consists of a two bar (called E and I bars) induction pick-off, a servo and a cam. During a straight and level flight through a uniform air mass the system is in balance and the pointer remains steady.

When a pressure change is met the movement of the capsules is transmitted mechanically (this being the only unassisted mechanical link in the whole system) to one of the two bars of the pick-off, moving the bar in response (fig. 11.7). This movement throws it out of balance and an error signal is raised in the pick-off. This signal is amplified and fed to the motor. As the motor turns it drives the cam which, in its turn, moves the other bar of the pick-off. When this bar has sufficiently moved to regain balance, the error signal disappears, movement of the motor, cam and the other bar stops and the system comes to rest.

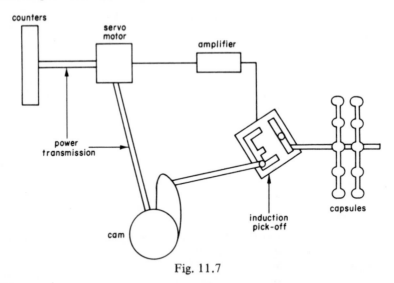

Fig. 11.7

As the motor turns, it also turns the height counters and the instrument's single pointer simultaneously with the cam. Thus, the transmission of information from capsules to counters can be said to be almost instantaneous.

Servo altimeters have the following advantages over conventional altimeters:
1.   At high altitudes very little pressure change takes place for a given change of altitude with the result that the capsule movement is considerably less than for the same change of altitude at lower levels. This factor makes ordinary altimeters inefficient at higher levels whereas the servo mechanism will pick up a capsule movement as little as 0.000 2 inches per thousand feet.

The pressure differences for 5 000 ft steps at low and high level are shown in this table:

Level (ft)	Pressure (mb)	Pressure difference (mb)
SL	1 013	—
5 000	843	170
10 000	697	146
55 000	91	—
60 000	72	19

At height, the capsule movement is very small, the pointer movement is relatively very large, causing an increased friction and inefficient operation in ordinary altimeters.

2.   Power transmission gives better accuracy.

3.   There is practically no time lag between arrival of new pressure and placing of the counters.

4.   Being an electrical system, correction for pressure error could be made and an Altitude Alerting device may be incorporated.

5.   Although conventional altimeters now employ digital presentation, it is generally more common with the servo altimeters. These eliminate the possibility of misreading.

6.   A pointer is still available — a useful asset at low levels in assessing rate of change of height.

The appearance of a typical indicator is shown in fig. 11.8.

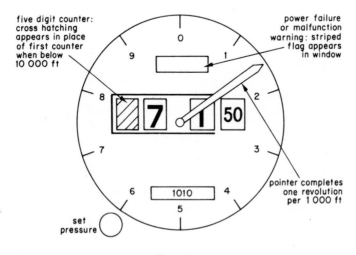

Fig. 11.8

12: Vertical Speed Indicator (VSI)

Vertical Speed Indicator (VSI) is a pressure gauge which utilises the principle of differential pressure to indicate an aircraft's rate of climb or descent.

Construction and operation

In construction, the VSI consists of a capsule held in an airtight case and fed with outside static pressure. Capillary delivery to the capsule is employed for the same reason as in the ASI: the tube is longer than in the ASI, as the VSI's capsule is more sensitive. Outside static is also fed to the inside of the case (that is, outside the capsule) but in this case it has to pass through a carefully calibrated restrictive device called the metering unit. The effect of the metering unit is to present static inside the case after a calibrated delay.

Thus, as the outside pressure alters, as it will when an aircraft commences to climb or descend, the capsule will be affected by the change almost immediately, whereas the change will reach inside the case only after a slight delay. This lag will occur every time the aircraft climbs or descends. Therefore as long as the aircraft maintains climb/descent attitude a differential will exist between the two (inside the case and inside the capsule). This differential will result in expansion or contraction of the face of

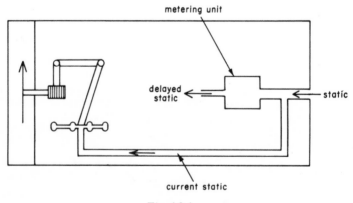

Fig. 12.1

the capsule depending on whether the pressure in the case is lower or higher than the pressure inside the capsule. For example, during a climb, denser pressure (appropriate to lower altitudes) will be present in the case and less dense (or current) pressure inside the capsule. The capsule will contract.

This expansion/contraction of the capsule is transmitted to the pointer system through a suitable mechanical link. The movement of the capsule being proportionate to the rate of change of pressure, the pointer indicates the rate of change of altitude. See fig. 12.1.

In level flight both pressures will be equal and the pointer will indicate 0 position.

In order to ensure the correct rate of flow through the metering unit in varying density conditions, a mechanical temperature/pressure compensatory device is generally introduced in the metering unit. Thus, correct flow is maintained during climb and descent.

Errors

1. Time Lag Error. During sudden changes in pitch attitude, a certain time must elapse before the new pressures reach the instrument and the differential is established.

2. Pressure Error. Any error in the static pressure being supplied to the instrument and caused due to its positioning on the fuselage will not affect the indications. This is because the instrument reads the differential pressure and not its magnitude. However, any disturbed airflow, caused by accelerations, decelerations, initiation of a climb or descent, lowering of the flaps, the undercarriage, will introduce an error. Errors induced by manoeuvres can cause any pressure instrument to misread for up to 3 seconds at low altitudes and up to 10 seconds at 30 000 feet. The times for the VSI are even longer than these (CAA Circular 3/1975). Thus, during any manoeuvres involving change of attitude or aerodynamic configuration, absolute reliance must not be placed on VSI.

3. Blocked static would give a zero indication.

The indicator

Straight and level position is indicated by the pointer at 9 o'clock position. The pointer travels through the top portion of the indicator (fig. 12.2) to indicate rate of

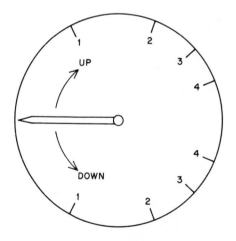

Fig. 12.2

climb up to 4 000 ft per minute. Similarly, it travels through the bottom portion of the indicator to indicate rate of descent up to 4 000 ft per minute. The maximum points are separated by an intervening space to prevent any chance of confusion as to which way the pointer movement took place. On some indicators the calibration is based on logarithmic scale (fig. 12.2), the largest pointer movement occurring near the zero position. This enables small variations from straight and level flight to be shown up.

13: Machmeter

Mach number (MN)

A machmeter indicates an aircraft's mach number. A mach number is defined as the ratio of the aircraft's true airspeed to the local speed of sound, that is, the speed of sound in the airspace in which the aircraft is flying. Expressed in a formula this is:

$$MN = \frac{V}{C}$$ where MN = mach number,
V = the aircraft's TAS,
C = local speed of sound.

Thus, a machmeter in fact gives the pilot a continuous indication of the ratio of his TAS to the local speed of sound.

It may be asked why this indication is necessary. Actually, this information is of vital importance to the pilot of high speed aircraft. As the flight approaches the speed of sound it is found that the behaviour of the aircraft changes. The physical principles involved at high speed flight are these.

As an aircraft travels through the atmosphere it creates vibratory pressure disturbances; these pressure waves travel ahead of the aircraft at the local speed of sound. They act to give an advance warning to the air ahead of the aircraft's approach which then diverges from the aircraft's path giving a streamlined flow past the aircraft. This is the normal operation. However, as the aircraft's speed increases this warning period gets shorter until the aircraft is flying at the speed of sound, at which time the air will give no warning at all.

Further, even at speeds well below the sonic speeds, say mach 0.75, the air over the thickest section of the wing could be travelling at sonic speeds. If this is so, then the physical conditions on the wing are the same as those which an aircraft flying at the speed of sound would have immediately on its front. These conditions give rise to a shock wave, and the pressure disturbances being created behind the wing will not be able to penetrate it.

The effect of this phenomenon is to increase the drag with possible loss of some lift. Violent impact with the air may cause buffeting and impose severe structural stresses on the airframe. An American test pilot described his sensations at this stage of the flight by comparing it with a motor cycle with iron wheels riding on a cobblestone road at full throttle.

As we noted, this condition does not just arise at the speed of sound: it arises well before this speed is reached. Also, the speed at which it occurs is not the aircraft's TAS, but how much notice it gives, by virtue of its speed, to the air ahead – this means, its speed in relation to the speed of sound. An aircraft which is designed for very high speeds generally employs very thin wing sections and the wings themselves are well swept back. These features delay the onset of the shock wave, and when it does occur, it is well in the rear.

That proportion of the speed of sound when the above described phenomenon occurs and a shock wave is formed for a given type of aircraft, is called 'critical mach number' or MCRIT. A pilot should not let his aircraft exceed this speed unless the aircraft is designed to fly beyond it. Thus, a knowledge of the mach number is of vital importance.

Principle of construction

We require the ratio of two elements — TAS and Local Speed of Sound. Now, TAS is the function of dynamic pressure, P — S, and the density. Speed of sound is the function of static pressure, S, and the density. Density being a factor to both sides of the fraction, the equation may be re-written as:

$$\text{Mach No} = \frac{P - S}{S}$$

P — S suggests ASI, or more appropriately an airspeed capsule. Similarly, S suggests altimeter capsule. Therefore, if we had two capsules, one responsive to airspeed and the other to altitude, placed 90° apart to give a ratio, by interlinking their movement to the pointer we could read off mach number. That is precisely what is done in a Machmeter (fig. 13.1).

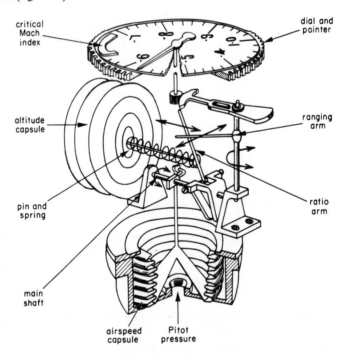

Fig. 13.1

Operation

Expansion/contraction of the airspeed capsule is transmitted through main shaft to the ratio arm. Similarly, expansion/contraction of altitude capsule is transmitted to the ratio arm through a pin which is kept in position by tension of a spring.

The airspeed capsule causes movement of the ratio arm in one plane, whereas the altitude capsule causes movement of the ratio arm in another plane. The two planes of movement are 90° apart.

The movement of the ratio arm is transmitted through the ranging arm to the pointer. The movement of the various arms is linked to the pointer in such a way that for an increased altitude or airspeed, a higher mach number is indicated.

Mach numbers are printed on the face of the instrument (fig. 13.2). An adjustable lubber mark is fitted over the dial as shown, to indicate critical mach number.

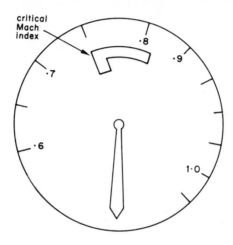

Fig. 13.2

Errors

This instrument only suffers from Instrument and Pressure errors, and these are similar to other pressure instruments.

Density Error. Machmeter does not suffer from this error since, as we saw above, density is a factor applicable to both sides of the equation and thus, its effect is eliminated.

Temperature Error. This again causes no error as it is eliminated with the density.

Compressibility Error. Since compressibility depends on $\dfrac{\text{dynamic}}{\text{static}}$ and the instrument is calibrated for this ratio, there is no compressibility error.

Therefore, since instrument and pressure errors are relatively small in value as compared with the TAS of the aircraft we can say that indicated mach number equals true mach number. In practice, mach number is always spoken of as indicated mach number and no distinction is made between Indicated, Rectified and True Mach Numbers.

A blocked static will affect the indications in all circumstances when the outside static changes. Thus, for example, if the static line gets blocked while the aircraft is descending, the machmeter will overread. This is because the altimeter capsule will not contract any further (as it ought to with increasing density at the lower levels), holding the pointer at too high a reading. The ASI capsule also works to the same end, since the old static offers less opposition to the pitot pressure. The overall effect is to give an overread. This effect may also be seen in the formula $MN = V/C$. As you are descending, the value of C is progressively increasing. But since you are using the value of C appropriate to the higher level, you are dividing V by too small a number, giving too high a mach number.

TAS
1.   In International Standard Atmosphere, RAS = TAS at mean sea level.
2.   As the height increases, for a given RAS, TAS increases and vice versa.
3.   RAS − TAS conversions may be made on the computer using the ISA rate of fall of temperature with height of 2° per thousand feet.

Speed of sound
1.   In the standard conditions the speed of sound is 661 kt at sea level and 573 kt at 36 090 ft and above.
2.   Thus, the speed of sound decreases as altitude is gained. The speed may be calculated from formula $C = 39\sqrt{T}$ where T is Absolute temperature. It may also be directly read off the computer: set MN index against corrected air temperature, and read off speed of sound on outer scale against 1 on the inner scale, or it may be interpolated − speed falling at 2.5 kt per thousand feet.

The mach number, speed of sound, TAS and RAS are all interrelated, and we will now study the effect of change of any one of these factors on the remainder.

Aircraft in a climb/descent
Mach number constant. Let us say that an aircraft is climbing from sea level to 36 000 feet (tropopause) at a constant mach number of 0.72. In doing so, the aircraft passes through the regions in which the value of the speed of sound, C, falls from 661 kt to 573 kt.

$$MN = \frac{V}{C}, \quad \therefore TAS = MN \times C$$
$$SL\ TAS = 0.72 \times 661 = 475.92\ kt$$
$$36\,000\ ft\ TAS = 0.72 \times 573 = 412.56\ kt$$

RAS at SL = TAS = 475.92 kt;
RAS at 36 000 ft for ISA temp. of −56.5°C = 230 kt (computer).

   Thus, in a climb at a constant mach number, the TAS decreases. RAS also decreases, but at a greater rate.
   In descent, the above set of figures reverses, that is, both RAS and TAS increase.
   A similar conclusion may be drawn from the inspection of the mach number formula and without use of any mathematics. In this formula, if you are climbing, the value of C decreases and if the mach number is kept constant, the TAS must also reduce proportionally.
TAS constant. If the TAS is kept constant, the RAS must decrease with altitude. In a climb, therefore, the RAS decreases; in a descent, the RAS increases.
   MN is the fraction TAS/C. As the aircraft climbs and the value of C decreases, the value of the fraction TAS/C increases and thus, the mach number increases. For similar reasons, in a descent the mach number decreases. Putting this in figures, a mach number of 0.75 at 36 000 feet gives a TAS of 429.75 kt. Descending with the

same TAS, at sea level, $MN = \dfrac{429.75}{661} = 0.65$, a decrease

RAS constant. Let us climb from sea level to 36 000 feet at a constant RAS of 240 kt.
At SL, RAS = 240 kt
     TAS = 240 kt

$$MN = \frac{240}{661} = 0.36$$

At 36 000 ft, RAS = 240 kt
          TAS = 420 kt   for ISA temp. of −56.5°C (computer)

$$MN = \frac{420}{573} = 0.73$$

Thus, in a climb at constant RAS, both the TAS and the mach number increase.
Examining the mach number formula, as the aircraft climbs at constant RAS, the
TAS must increase and if the TAS increases then the value of the fraction TAS/C
increases. This is the increase in the mach number indication. At the same time, as
the aircraft climbs, the value of C decreases, increasing the value of the fraction
and the mach number.

In a descent, the TAS decreases, and so does the mach number.

Aircraft in a level flight
Even when the RAS or mach number of the aircraft are unaltered on purpose, the
mach number indication may change because of changes in the outside temperature.
When flying from regions of high temperature to regions of low temperature, the
speed of sound decreases. If the mach number is to be kept constant, the TAS must
proportionally decrease. Similarly, when flying from cold air to warm air for
constant mach number, the TAS must increase.

For the above reasoning, if two aircraft are flying at different levels but doing the
same mach number, the one at the lower level has higher TAS.

Serviceability checks
On the ground: check that the pointer is not resting at any place other than at 0
position.

In flight: a rough check may be made by noting a reading of .5 mach against IAS
of 330 kt near sea level.

Accuracy
±0.01 mach, except at limit of the range where it falls to ±0.02 M.

Machmeter problems
1.   The TAS of an aircraft flying in the atmosphere where the speed of sound is
600 nm per hour is 450 kt. What is its mach number?

$$MN = \frac{V}{C} = \frac{450}{600} = 0.75 \quad \text{Answer}$$

2.   An aircraft is doing a mach number of 0.7 whereas its TAS is 430 kt. What is the
speed of sound?

$$C = \frac{V}{MN} = \frac{430}{0.7} \text{ kt} = 614.3 \text{ kt} \quad \text{Answer}$$

3.   An aircraft is doing mach 0.66 in an atmosphere where the speed of sound is 615 kt. What is its TAS?

$$V = MN \times C$$
$$= (0.66 \times 615) \text{ kt}$$
$$= 405.9 \text{ kt}\quad \text{Answer}$$

4.   A computer may be used in this problem. The rectified airspeed of an aircraft at FL 350 is 230 kt. The corrected outside air temperature is −60°C. (a) What is the local speed of sound? (b) What is the aircraft's mach number?

   (a) Set the mach number index in the airspeed window against −60°C on the corrected air temperature scale. Read off the speed of sound on the outside scale against 1 on the inside scale. Answer: 568 kt.

   (b) RAS converted to TAS is 392 kt (density and compressibility errors). Set the speed of sound against 1 on the inside scale and read off the mach number on the inside scale against TAS 392 kt on the outside scale. Answer: 0.69.
(Note: different computers read slightly different answers.)

5.   An aircraft is doing mach 0.82 in an atmosphere where the speed of sound is 1 060 ft/s. What is its TAS?
First convert the speed of sound from feet per second to nm per hour.

$$\text{Speed in kt} = \frac{1\,060 \times 3\,600}{6\,080}$$
$$= 628 \text{ kt}\quad \text{(computer)}$$

(Note. Alternatively, the relationship 60 mph = 88 ft per second may be used but do not forget to convert mph to kt.)

$$TAS = MN \times C$$
$$= 0.82 \times 628$$
$$= 515 \text{ kt.}$$

6.   If mach 1 at sea level equals 740 mph, give the speed of sound in feet per second.
In one hour, distance = 740 sm
$$= 740 \times 5\,280 \text{ ft}$$
and in one second, distance $= \dfrac{740 \times 5\,280}{3\,600}$ ft
$$= 1\,085 \text{ ft}\quad \text{(computer)}$$
Speed of sound is 1 085 ft/s.

7.   At FL 400 where the temperature deviation from standard is +10°C, a TAS of 470 kt is obtained when operating at mach 0.82. What temperature deviation is required to obtain the same TAS for the same mach number when flying at FL 320?

   Mach number/TAS relationship only depends on the local speed of sound and is independent of the flight level being flown. The speed of sound, in its turn, is dependent on the ambient temperature which means that for every temperature there is one speed of sound. Thus the problem involves first discovering the actual temperature at FL 400.

FL 400:   Standard temp. = −65°C
                       Deviation   = +10°C
                       Actual   = −55°C

What we have now established is that the speed of sound that results from an ambient temperature of −55°C gives a TAS of 470 kt for mach 0.82.

FL 320:   Standard temp.   = −49°C
                       Temperature necessary = −55°C
                       Deviation necessary   = − 6°C   Answer

8. If the local speed of sound is 450 kt, what will be the increase in TAS if the mach number is increased by 0.06?

For a constant speed of sound, the TAS increases in direct proportion to the increase in mach number. Therefore

$$\text{increase in TAS} = (450 \times 0.06)\,\text{kt}$$
$$= 27\,\text{kt}\quad\text{Answer}$$

9. Temperature in the jet standard atmosphere decreases at 2°/1 000 ft from +15°C at sea level with no tropopause. The speed of sound is 660 kt at +15°C and varies as the square root of the absolute temperature.

Using these assumptions, give the temperature (°C) in the jet standard atmosphere where the TAS is 400 kt for a mach number of 0.735.

For a TAS of 400 kt at mach 0.735, the speed of sound is 545 kt (computer). Therefore, the ratios are:

$$\frac{545}{660} = \frac{\sqrt{T}}{\sqrt{(288)}};\quad \sqrt{T} = \frac{545\sqrt{(288)}}{660}\quad\text{and}\quad T = \frac{545^2 \times 288}{660^2}$$

$$= 196.4\quad\text{(by log tables).}$$

where T is the absolute temperature and A = 273 + °C

∴ 196.4 − 273 = −76.6°C.

Alternatively, on the computer, set TAS on the outside scale against mach number on the inside scale. Against 1 on the inside scale, read off the local speed of sound. Then solve the formula $C = 39\sqrt{T}$.

10. At FL 280 a decrease in mach 0.1 results in a decrease in TAS of 58 kt. What is the local speed of sound?

$$C = \frac{V}{MN} = \frac{58}{0.1}\,\text{kt} = 580\,\text{kt}\quad\text{Answer}$$

14: Gyroscope

Try balancing a stationary bicycle wheel and see how hard the task is. Now give it a gentle tap at the rear and make it roll. You will notice that while it is rolling at reasonable speed it does not fall down. The explanation is contained in the gyroscopic properties of the rolling wheel. A gyroscope may be described simply — any wheel that spins on its axis is a gyroscope. A spinning top, motor car wheels, aircraft propellers — these are just a few examples.

When a wheel spins on its axis it acquires two properties called Rigidity in Space and Precession.

Rigidity is the reluctance of a gyroscope to change the direction of its spin axis. This property is acquired in obedience to Newton's First Law of Motion which states that a body continues to remain in state of rest or uniform motion unless an external force is applied to change that state.

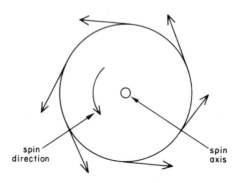

spin
direction

spin
axis

Fig. 14.1

In fig. 14.1 the gyro axis is in the horizontal plane and the wheel is spinning in the vertical plane. As the wheel spins, each particle of the wheel tends to continue in its instantaneous direction as shown in the figure. This is to be expected, according to Newton. The fact that the particles do not fly off the wheel due to the tensile strength of the metal does not in any way affect this directional characteristic acquired by the moving particles. Any attempt to change the direction of the spin axis in any plane will therefore be met with opposition. This is the property of rigidity: once spinning, the gyro axis will remain pointing in the same direction in space. For the bicycle wheel to fall down while rolling, the spin axis must change its direction by 90° and unless external force is applied, it will not do so as long as it has a reasonable speed. The magnitude of rigidity is directly proportional to the speed of the wheel. The faster it spins, the greater the rigidity it acquires. It is also proportional to the moment of inertia of the wheel. To improve moment of inertia, the gyro wheel is given as large a radius as the design factors would permit and the bulk of its weight is concentrated at the rim. If you think back to the magnetic compass you will notice that this is just the opposite of what you did in that compass to achieve aperiodicity. Finally, rigidity is inversely proportional to the external force applied. To put this in mathematics,

$$R \propto \frac{SI}{F}$$

where R = rigidity
S = speed of the wheel
I = moment of inertia and
F = External Force

Precession

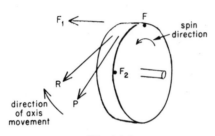

Fig. 14.2

In fig. 14.2, P is the instantaneous direction of travel of a moving particle in a wheel spinning with its axis horizontal. F is a force applied at top of the wheel, resulting in force vector, F_1. If the wheel was stationary the effect of F_1 would be to topple the gyro, since vector P is absent. In the case of a spinning wheel, however, the two vectors will resolve into a resultant direction, R, and the axis will move clockwise from P to R. This is like saying that force was applied at point F_2 on a stationary wheel. This is precession.

Fig. 14.3

In fig. 14.3(a) force F is applied so as to rotate the axis anti-clockwise in the horizontal plane, as viewed from above. (Force F is applied at the paper, coming out towards us.) That force will travel through 90° and act at point F_1 and the gyro will topple — fig. 14.3(b). This is again precession. Precession can, therefore, be described as follows: when a force is applied to a spinning wheel, that force does not act at the point of application, but acts at a point which is 90° removed from it in the direction of the spin. Going back to the bicycle wheel, if you wish to turn the direction, you tap the appropriate side of the wheel at the top. A stationary wheel would have fallen down with this tap. In a moving wheel this force travels forward through 90° and acts. Thus, say the tap was from left to right. This left to right force acts at the forward-most point (90° removed in the direction of the spin) of the wheel and turns the wheel to the right.

In order to precess a gyro we must overcome its rigidity. The higher the speed, the greater the force required to precess it. Similarly, the larger the moment of inertia the greater the force required. Mathematically, therefore:

$$\text{Precession} \propto \frac{F}{SI}$$

These two expressions reveal that rigidity and precession are opposing terms, which is quite true. Rigidity is the reluctance to move; precession is the tendency to move.

Types of gyroscope

The basic classification of gyroscopes is as follows:

Space Gyro. A gyroscope having freedom in all three planes is called a space gyro. The three planes relate to the three axes of aircraft, i.e. fore-and-aft axis, athwartships axis and vertical axis. There is no means of any external control over a space gyro, a feature which distinguishes it from tied gyro and earth gyro.

Tied Gyro. This is a space gyro which has a means of external control. Being basically a space gyro a tied gyro has freedom in all three planes. An uncontrolled space gyro would be of no practical use in an aircraft instrument where the gyro is required to be set up and maintained in a certain direction. We will learn the control systems with individual instruments later.

Earth Gyro. This is, again, a space gyro but controlled by the gravity of the Earth.

Rate Gyro. This is a gyro having one plane of freedom only, the plane of freedom being 90° removed from the plane of rotation. It is utilised to measure rate of turn and we will learn more about it later.

Gyro wander

Any deviation of the gyro axis from its set direction is termed as gyro wander. It is of two types.

Real Wander. Any physical deviation of the gyro axis is called real wander. A gyro axis ought not to wander away but various forces do effectively act on the gyro and cause it to precess. For example, some friction is always present at the spin axis. If this friction is symmetrical it merely slows down the rotor; if it is asymmetric, a force will arise which will have a precessing effect on the gyro. Similarly, friction in the gimbal rings (rings that hold the rotor and allow it movement in various planes) will precess the gyro. Another instance is the shift of the centre of gravity of the gyroscope from its dead centre position. This usually occurs as a result of wear on the gyro. When flying through turbulent air, turbulent vectors acting on the gyro will resolve in an ultimate direction and a precessing force will result. These errors are not constant; they vary with time and therefore no calibration cards are produced.

Apparent Wander. As the term suggests the gyro axis in this case does not physically wander away and yet to an observer it appears to have changed its direction. The reason for this is quite simple: the gyro maintains its direction with reference to a fixed point in space whereas we, on the Earth, rotate with the rotating Earth. Thus, since we ourselves do not maintain a fixed direction in space, the gyro must appear to us with the passage of time to have altered direction. We will study the effects due to the Earth's rotation on a horizontal axis gyro and a vertical axis gyro separately.

Horizontal axis gyro

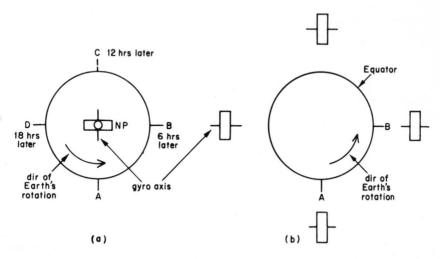

Fig. 14.4

In fig. 14.4(a) we have a horizontal axis gyro at the North Pole. An observer at A sets the gyro so that its axis is aligned with him, as shown in the figure. Six hours later the Earth will have rotated through 90° and the observer will be viewing the same gyro from position B. To the observer who does not appreciate his own velocity, the gyro axis appears to have moved clockwise in the horizontal plane through 90°. Twelve hours later he will be observing the gyro along its axis but from the opposite end. Eighteen hours later he will be at position D and the axis will appear to have rotated through 270°. Finally, twenty-four hours later he will be again at Position A and view the axis just as he had initially set it.

This is an example of apparent wander. Apparent wander in the horizontal plane as we discussed above is known as gyro drift, and since at the Poles there is a drift of 360° in 24 hours, we can say that the drift experienced at the poles is the maximum possible value (that is, rate of drift is the same as the angular velocity of the Earth). To be more precise the rate of drift is 15.04° per hour.

In the above discussion, you would notice that with the horizontal axis gyro all the apparent movement occurs only in the horizontal plane. In other words, the gyro does not show up any movement in the vertical plane in terms of its axis rising or falling. Any movement of the gyro axis in the vertical plane is called gyro topple, and in this case there is no topple.

In fig. 14.4(b) the gyro is a horizontal axis gyro and is placed at the Equator. Here, its axis may be aligned in the north-south direction, that is along the observer's meridian or it may be aligned in the east-west direction, that is along the Equator. At the Equator all meridians are parallel to each other and a gyro axis aligned with a meridian will remain with that meridian throughout the period of twenty-four hours. This means that there will be no drift and no topple. If the axis is aligned with the Equator as in the figure it will be noticed that six hours later the horizontal axis gyro will appear to be a vertical axis gyro (position B). This is an apparent change in the vertical plane and therefore, a topple. Further, the axis appeared to move through 90° in six hours, which means the rate of topple at the Equator is the maximum possible (15.04° per hour). The axis otherwise does not appear to move in the horizontal plane and thus there is no drift. To summarise:

Horizontal axis gyro: at Poles   : maximum drift; no topple;
                    at Equator : no drift; maximum topple.

Vertical axis gyro

In fig. 14.5(a) an observer at the North Pole sets up a gyro with its axis vertical. It will be apparent that the Earth will rotate beneath the axis, and the gyro axis will neither appear to topple nor drift.

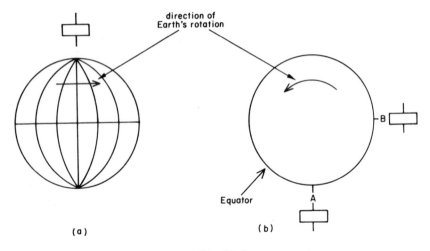

Fig. 14.5

In fig. 14.5(b) the observer is shown at the Equator at position A with gyro axis vertical. The effect is similar to the horizontal gyro in fig. 14.4(b). Six hours later when in position B the gyro axis will appear to have spun through 90° in the vertical plane to become apparently a horizontal axis gyro. To summarise:—

      Vertical axis gyro: at Poles   : no drift; no topple;
                        at Equator : no drift; maximum topple.

Combining the two we may summarise that drift at the Poles is 15.04° per hour and 0 at the Equator. Topple is 0 at the Poles and 15.04° per hour at the Equator. From this, drift and topple at intermediate latitudes bear the following relationship:

      Drift   = 15.04° x sine of the latitude per hour, and
      Topple = 15.04° x cosine of the latitude per hour.

Let us take an example to clarify the discussion. Say we set up a horizontal axis gyro, the axis facing east-west direction at 30° N. We want to know what will be the attitude of the axis after four hours. Since it is a horizontal axis gyro at an inter-mediate latitude facing east-west, it will both drift and topple.

Drift = 4 x 15.04 sin 30
     = 30.08°

From fig. 4(a) we note that the gyro axis drifts clockwise in the northern hemi-sphere (and therefore anti-clockwise in the southern hemisphere). Thus, the axis is now aligned with 090 + 30.08 = 120.08°/300.08° direction.

Topple = 4 x 15.04 cos 30
      = 52.1°

As the Earth travels from west to east and the axis remains pointing in the same direction in space, the eastern end of the axis will appear to have risen by 52.1° and the western end to have been similarly depressed.

15: Directional Gyro (DGI)

Directional Gyro Indicator (DGI) employs a horizontal axis gyro and utilises the principle of rigidity to indicate aircraft's heading. A spinning rotor, as we saw in the previous chapter, has the property of rigidity in space. Thus once the DGI rotor attains its full speed and its axis is manually aligned with a datum (true or magnetic north) it will continue to point in that direction in space during the rest of the flight. When the aircraft alters its heading it does so relative to the gyro axis, that is, the aircraft and the gyro case turn about the gyro axis. Changes in headings are thus indicated instantaneously.

It also utilises the property of precession for two purposes:
(a) to provide gyro control and (b) to compensate for apparent wander.

Construction

The DGI rotor is mounted in two rings called the inner gimbal and outer gimbal. Each gimbal has movement independent of the other. The rotor is mounted in the inner gimbal, the gimbal itself lies in the horizontal plane and the rotor which it holds spins in the vertical plane. The inner gimbal is mounted in the outer gimbal pivoted at two points on the inner gimbal which are 90° removed from the rotor axis. See fig. 15.1.

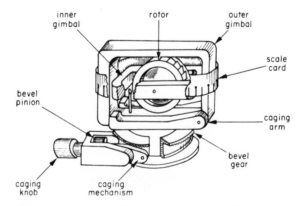

Fig. 15.1

This allows the rotor to move about the horizontal axis. This movement, however, is restricted to 110°, that is, 55° either side of the rotor's vertical plane. This limits the aircraft's manoeuvres in pitch and roll and if the limits are exceeded the inner gimbal will come in contact with a mechanical stop and the gyro will precess. The restriction is necessary in order to prevent the inner gimbal from coming in contact with the outer gimbal and damaging the instrument.

The outer gimbal is mounted in the case of the instrument and pivots about the vertical plane. This ring has freedom of 360° and carries the scale card. The rotor is driven by a jet of air from a nozzle, air entering the rotor case through a hole in the periphery of the case. Air impinges on small buckets carved out on the rim and spins the rotor at about 12 000 rpm.

The initial setting of the heading and subsequent resettings during the flight are carried out by use of the gyro caging control. The caging knob on the instrument is depressed and turned until the required heading appears in the window. When the knob is so depressed a bevel pinion engages a system of bevel gear, part of the outer gimbal. Thus, by turning the knob, the outer gimbal carrying the scale card is turned. The knob, when depressed, also locks the inner gimbal by means of a caging arm. This prevents the rotor from toppling while the force is being exerted on the outer gimbal. (Set the heading when the aircraft is flying straight and level as otherwise a small error may occur in the reading.)

The gyro should be caged during violent manoeuvres or manoeuvres likely to exceed the limits. The caging control is also used to erect the gyro should it inadvertently topple. This is because the caging arm rights the rotor before it locks it.

Finally it is important to remember that the gyro must be uncaged and re-synchronised before the DGI is used again.

Gyro control

When an aircraft alters its heading it turns about its vertical axis. Thus, the heading measurements will only be correct when measured with reference to that axis. This means that the rotor attitude must be controlled so that during banked turns it remains spinning in the aircraft's vertical plane (and not true vertical). This is achieved as follows. In fig. 15.2(a) the aircraft is straight and level, the rotor is in the aircraft's

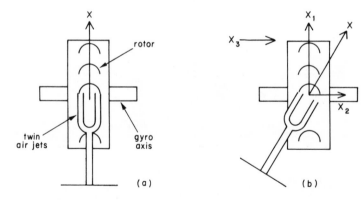

Fig. 15.2

vertical plane and the air jets driving force, X, is fully utilised to drive the rotor. In fig. 15.2(b) the aircraft is in right bank. It will be seen that the driving force is not fully utilised to drive the buckets. Therefore, this force, X, must break down into two components, X_1 and X_2 at 90° to each other. X_1 acts along the rotor, now driving it at reduced speed. X_2 acts on the rim of the rotor. This component will precess through 90° in accordance with the precession rule in the direction of spin and act at X_3 in the direction shown. The effect of this force is to precess the rotor until it is aligned once again with the air jets as illustrated in fig. 15.2(a). The rotor is then in the aircraft's vertical axis.

Gyro drift

All horizontal axis gyroscopes are subject to drift as explained in the previous chapter, and the DGI is no exception. The drift occurs due to both real and apparent wander and the DGI should be reset at regular intervals during the flight. An instrument in good condition should not precess more than 4° every 15 minutes.

The real wander occurs due to friction, static unbalance (e.g. shift of the centre of gravity) and air turbulence. There is no means of predicting these errors to enable us to allow for them since wear and tear alters the values. The apparent wander occurs at the rate of approximately 15.04° sin lat per hour. The effect of this on heading indications is discussed in the following paragraphs.

We will take the northern hemisphere for illustration and study the effect on the heading reading in four situations, that is, effect when aircraft is stationary, aircraft on easterly track, aircraft on westerly track and finally, aircraft on northerly-southerly track.

In fig. 15.3(a) below let us say that we lined up the DGI with the local meridian at position A. The heading indicated is 090°. Consider the effect one hour later, the

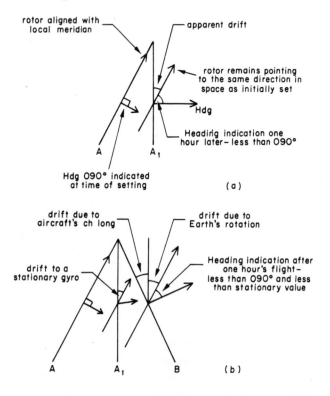

Fig. 15.3

gyro remaining stationary. The local meridian will have changed its direction in space due to the Earth's rotation and let us say that it arrives at position A_1. Now, the gyro rotor remains aligned with the local meridian direction one hour ago (assuming that there is no real drift) and this phenomenon must cause a discrepancy between the two gyro readings, that is the initial reading of 090° and the reading one hour later. The error occurs at the rate of 15.04 sin lat/hr and the readings, as will be seen from the diagram, decrease.

Suppose the aircraft took off from A on an easterly track. One hour later, when the Earth's meridian A has reached A_1, the aircraft will have travelled to a new meridian, B. At B the heading will be in error due to two causes: Earth's rotation and the aircraft's change of longitude (again assuming no real wander) and the reading will be still less, fig. 15.3(b).

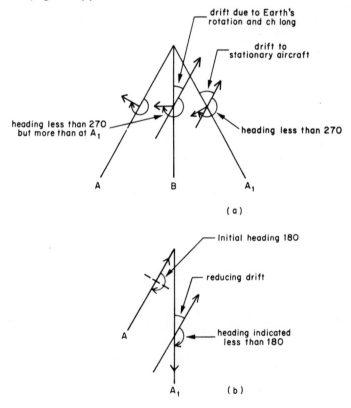

Fig. 15.4

In fig. 15.4(a) the gyro is shown with its axis aligned with local meridian A and the aircraft is on a track of 270°. In one hour's time meridian A will have changed direction to A_1 and if the aircraft was stationary an error of 15.04 sin lat would be recorded, the readings reducing. However, the aircraft flying on westerly track will arrive at longitude B in one hour's time. The reading at B will still be less than 270° but more than what would be indicated if it remained stationary. The aircraft in the illustration is presumed to have its speed less than the speed of the Earth at the local parallel. It will be appreciated that if an aircraft is flying at the same speed as the speed of the Earth but on a westerly track it would negative the Earth's easterly velocity. In that case it would reach position A in the same time that original A displaced itself to A_1 and the gyro would give no error.

In fig. 15.4(b) the aircraft having aligned its gyro to the local meridian sets off from A on a southerly track and arrives at A_1 after one hour. The aircraft is flying down the same meridian; the deviation to A_1 is due to the Earth's rotation. The reading here is less than 180, and it should be noticed that the magnitude of the error is a continuously decreasing value as the aircraft travels southward (sine of Equator is 0). Similarly on northerly track the readings decrease but the rate of decrease quickens as the flight progresses northward.

Southern Hemisphere. The results are opposite to NH in that the readings increase. One example is sufficient to show this. In fig. 15.5 an aircraft on longitude A sets up its gyro with the local meridian and the reading is 090°. If the aircraft remains

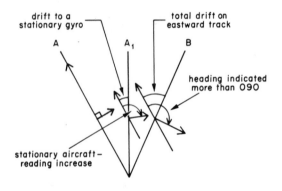

Fig. 15.5

stationary, in one hour's time meridian A will change direction to A_1. The indicator reading as can be seen in the figure has increased, the rate of increment still being 15.04° sin lat. If the aircraft was flying on an easterly track it will arrive at B in one hour and the heading indicated will have increased still further.

To summarise – unbalanced gyro

Northern hemisphere

Aircraft travelling east   – the readings decrease.
Aircraft travelling west   – the readings decrease but at a lesser rate.
Aircraft travelling south   – the readings decrease at a decreasing rate.
Aircraft travelling north   – the readings decrease at an increasing rate.

Southern hemisphere

Aircraft travelling east   – the readings increase.
Aircraft travelling west   – the readings increase at a lesser rate.
Aircraft travelling south   – the readings increase at an increasing rate.
Aircraft travelling north   – the readings increase at a decreasing rate.

Compensation for apparent wander

Where an aircraft operates in the vicinity of a given latitude the gyroscope may be mechanically precessed equally and in the opposite direction to the drift experienced at that latitude, thus compensating for the apparent wander. The mechanism involved is a nut called the latitude nut which is screwed on the horizontal axis of the gyroscope. A balance weight at the opposite end of the axis balances the weight of the nut when the nut is riding in the central position. The latitude nut is then screwed inward or outward as required to produce a force in the vertical plane which precesses the rotor in the horizontal plane. In fig. 15.6:

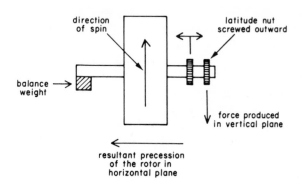

Fig. 15.6

the nut is screwed outward to compensate for drift in the southern hemisphere. It should, however, be appreciated that it only compensates for that latitude for which it is corrected, as long as the gyro remains stationary. Errors will arise on all tracks, and in assessing these errors the effect of the compensating device will have to be considered. A latitude balanced gyro will give errors as follows:

N. hemisphere (the correction is +15° sin lat per hour to compensate for a decrease)

Stationary:	nil drift at the latitude of compensation
Fly east:	readings *decrease* as aircraft speed adds to earth's rotation, so 15° sin lat per hour is too small a correction
Fly west:	readings *increase* as aircraft speed subtracts from earth's rotation, so 15° sin lat per hour is too large a correction
Fly north:	readings *decrease:* sin lat is getting larger, and 15° sin lat per hour is insufficient correction
Fly south:	readings *increase:* sin lat is reducing in value, and a correction of 15° sin lat is too large

S. hemisphere (the correction is −15° sin lat per hour)

Stationary:	nil drift at the latitude of compensation
Fly east:	readings *increase* as aircraft speed adds to speed of earth's rotation, so correction of 15° sin lat per hour is not enough
Fly west:	readings *decrease* as aircraft speed subtracts from speed of earth's rotation, so correction of 15° sin lat per hour is too much
Fly north:	readings *decrease* as sin lat decreases, so correction of 15° sin lat is too much

Fly south: readings *increase* as sin lat gets greater, and correction of 15° sin lat is not enough.

Finally on this important point, if a gyro is compensated for wander at 60°N, an amount of drift of +15° sin 60° has been put in. Taking that gyro to 30°N means that too much has been put in and the readings would be in error on this account only. Taking the same compensated gyro to 60°S would mean that +15° sin 60° has been put in to add still to the 15° sin 60° wander there anyway − in fact, doubling the drift.

Gimballing errors

These are minor errors which occur due to the geometry of the gimballing system. They occur when the aircraft is banked, in a climb or a descent. They disappear as soon as straight and level flight is resumed.

Type of gyro

Gyro used in DGI is a tied gyro, having freedom in three planes. The axis of the gyroscope is aligned with the 000 − 180 indication. Being a horizontal axis gyro, the forces imposed on the axis during a turn result in the movement of the rotor in the vertical plane. Any such movement has no effect on the heading indication and therefore a DGI does not suffer from conventional turn and acceleration errors.

Gyro problems

The gyro problems revolve about four basic conditions: the gyro is unbalanced; the gyro is balanced for a latitude; the gyro is stationary; the gyro is moving. All problems will have somewhere in them the application of the fundamental formula: drift = 15° sin lat. Repeating here, the readings decrease in the northern hemisphere; increase in the southern hemisphere. This is purely the effect of the earth's rotation and we are not concerned at this stage whether the gyro is balanced or unbalanced, stationary or mobile. When the gyro is mobile, the changing meridians modify the result in the formula 15° sin lat. When the aircraft is on an easterly track the amount of convergency (ch long x sin mean lat) is added; when on a westerly track it is subtracted from the result of 15° sin lat. And this rule applies to either hemisphere. We now propose that you adopt the following layout in working out your gyro problems.

1.   **Apparent wander due to the earth's rotation (15° sin lat)** _____
                                        (enter decrease if in the NH; increase if in the SH)

2.   **Apparent wander due to convergency (ch long x sin lat)** _____
                                        (+ if easterly Tr; − if westerly Tr)

3.   **Sum of items 1 and 2 is the total wander**   _____   **(inc or dec)**

4.   **Add the real wander (or lat nut)**   _____   **(inc or dec)**

5.   **The sum is the total drift**   _____   **(inc or dec)**

Note that you may not need to use all the steps in the above layout, neither would you always start at the top and work downwards. You may start at the bottom and work upwards. The idea is to complete initially as much of your proforma as possible from the data in the question and work out the rest. You will also notice that when the question involves the aircraft's new meridian, or the distance travelled in an E/W direction (departure) or the aircraft's ground speed, the information in item 2 will yield the answer. All these factors are directly concerned with the convergency.

$$\text{Convergency} = \text{ch long} \times \sin \text{mean lat.}$$

Ch long x sin mean lat is the figure that goes in item 2. If departure is given in the question you may convert it to convergency in two steps.

(a) Ch long = departure x sec lat.

(b) Convergency = ch long x sin mean lat.

These two steps may be cut down to a single simple operation as follows:

$$\text{ch long} = \frac{\text{dep}}{\cos \text{lat}} \quad \text{and} \quad \text{convergency} = \text{ch long} \times \sin \text{lat}$$

$$\therefore \text{convergency} = \frac{\text{dep}}{\cos \text{lat}} \times \sin \text{lat}$$

$$= \text{dep} \times \tan \text{lat}$$

and since we want convergency in degrees, the departure in the above formula is also in degrees of longitude.

The same formula may be rewritten to find departure:

$$\text{departure} = \frac{\text{convergency}}{\tan \text{lat}}$$

and since departure is in nautical miles rather than in degrees of longitude, the convergency in this formula is in minutes of longitude. All this should be crystal clear after you have done a few problems.

1. An unbalanced gyro free from real wander is set up at 60°N. What will be the effect on the gyro reading after one hour?

Solution

$$\begin{aligned}
\text{Apparent wander} &= 15° \sin 60° \\
&= 15° \times 0.866 \\
&= 13° \quad \text{reading's decrease in one hour.}
\end{aligned}$$

2. A directional gyro is balanced at 45°S. If it is free from other real wander, what will be the effect on the reading if the gyro remains stationary at 45°S?

Answer: The gyro will continue to give correct readings.

3. A directional gyro is balanced so that the drift is zero when stationary at 45°S. What drift will it experience when stationary at 60°N assuming that there is no real wander apart from that due to the lat nut?

Solution

The first step is to ascertain how much correction has been applied at 45°S. This correction will be equal in magnitude and opposite in sign to the apparent wander at 45°S.

$$\begin{aligned}
\text{Gyro drift at } 45°S &= 15° \sin 45° \\
&= 15° \times 0.7071 \\
&= 10.6°/\text{hr}
\end{aligned}$$

and since this is southern hemisphere, the readings increase. Therefore, the real wander due to compensatory measures is 10.6°/h decrease. Now we can work out the effect at 60°N.

App. drift at 60°N = 15° sin 60° = 13°/h decrease
Real drift   = 10.6°/h decrease (lat nut)
Total   = 23.6°/h decrease

4.   A perfect directional gyro has been balanced for 45°S. What drift will it experience when flying along the parallel of 45°S on a track of 090°(T), with ground speed 240 kt?

Solution
App. wander at 45°S = 15° sin 45°   = 10.6°/h increase

Convergency = dep tan lat = $\dfrac{240}{60}$ tan 45°   = +4.0°/h (easterly)

Total   = 14.6°/h increase
Real wander = 10.6°/h decrease
Total drift   = 4.0°/h increase

5.   A perfect gyro is balanced to give zero drift when stationary at 45°S. What drift will it experience when flying on a track of 270°(T) along 45°S at a ground speed of 240 kt?

Answer: 4°/h decrease.

Solution
App. wander 45°S   = 15° sin 45°   = 10.6°/h inc
Conv   = dep tan lat   = 4 tan 45°   = −4° /h (west)
Total   6.6°/h inc
Real (lat nut)   10.6°/h dec
Total   4.0°/h decrease

6.   A gyro is compensated for 45°N. What drift will be experienced when flying on 090°(T), at ground speed 240 kt in latitude 60°N?

Answer: 9.3°/h decrease.

Solution
Correction at 45°N = 10.6 in.
App. wander at 60°N   = 15° sin 60°   = 13.0°/h dec
Conv = dep tan lat   = 4 x 1.7321   = +6.9°/h (east)
Total   19.9°/h dec
Real wander (lat nut)   10.6°/h inc
Total   9.3°/h dec

7.   A directional gyro is adjusted to give zero drift at 50°N. What drift will be experienced when flying at 60°N on a track of 270°(T) at a ground speed of 360 kt?

Answer: 8.9°/h increase.

Solution

App. wander at 50°N = 15° sin 50° = 11.5°/h dec
∴   Correction = 11.5°/h inc

App. wander lat 60°N = $\qquad$ 15 sin 60° = 13.0°/h dec
Conv = dep tan 60° $\qquad$ = 6 tan 60   −10.4°/h (west)

	Total	2.6°/h dec
	Real (lat nut)	11.5°/h inc
	Total	8.9°/h inc

8.   A stationary gyro on being tested at 58°N gives a drift of 9°/h, the readings decreasing. At which latitude would this gyro give zero drift?

Answer: 14°17'N.

Solution

The first thing we want to check is the apparent wander at 58°N to see if it is in fact 9°/h decrease. If it isn't then there is some real drift which we must calculate.

App. drift at 58°N = 15° sin 58° = 12.7°/h dec
Actual drift in the gyro $\qquad$ = 9.0°/h dec
∴ Real drift $\qquad$ = 3.7°/h inc

We know now there is real drift in the gyro which increases the reading at the rate of 3.7°/h. In order to reduce this drift to zero all we have to do is to take the gyro to a latitude where the apparent drift is the same in magnitude as the real drift, but opposite in sign. Thus, in the present problem the real drift is 3.7°/h, increase: the latitude we want is where the apparent wander is 3.7°/h, decrease. The hemisphere is immediately indicated by the term decrease: the latitude we are looking for is in the northern hemisphere. Now for the second part

$$\text{app. drift} = 15° \sin \text{lat, or}$$

$$\sin \text{lat} = \frac{\text{app drift}}{15°}$$

$$= \frac{3.7°}{15°}$$

$$= 14°17'N$$

9.   With reference to problem 8 above, what drift would you expect at 30°S?

Answer: 11.2°/h increase.

Solution

App. wander at 30°S = 15° sin 30°	= 7.5°/h inc
Real drift	= 3.7°/h inc
Total	11.2°/h inc

10. A gyro on a ground test at 60°S is found to have a drift of 8°/h, the readings increasing. Where would this gyro give zero drift?

Answer: 19°28'S.

Solution

App. wander at 60°S = 15° sin 60°	= 13°/hr inc
Actual drift	= 8°/hr inc
Real drift	= 5°/hr dec

$$\sin \text{lat} = \frac{\text{drift}}{15°}$$

$$= \frac{5°}{15°}$$

$$= 19°28'S$$

11. On ground test a gyro is found to have a drift of 6°/h at 30°N, the readings decreasing. What drift would it experience when flying on a track of 270°(T) at 50°S if the aircraft's ground speed is 360 kt?

Answer: 5.8°/h increase.

Solution

App. wander at 30°N = 15° sin 30° = 7.5°/h dec
            Actual drift   = 6.0°/h dec
            Real drift   = 1.5°/h inc
At 50°S app. wander = 15° sin 50° = 11.5°/h inc
Conv = dep tan lat = 6 tan 50°   = −7.2°/h (west)
            Total   4.3°/h inc
            Real drift   = 1.5°/h inc
            Total drift   = 5.8°/h inc

12. A DGI tested at 50°S gives a drift of 6°/h, the readings increasing. In what direction and at what speed must this aircraft fly at 30°S for the gyro drift to be zero?

Answer: Direction west (270°T); GS 207.9 kt.

Solution

App. wander at 50°S = 15° sin 50° = 11.5°/h inc
            Actual drift   = 6.0°/h inc
            Real drift   = 5.5°/h dec
In our gyro problem lay out
1.   App. wander 30°S = 15° sin 30°   _____
2.   Conv = dep tan lat   _____
3.   Total app. wander   _____
4.   Real wander   _____
5.   Total drift   _____
we know the last item, that is, the total drift which is zero.
Enter 0 in the above blank layout and work backward.
App. wander at 50°S = 15° sin 50° =   = 11.5°/h inc
Actual wander is   = 6.0°/h inc
∴ Real drift present   = 5.5°/h dec
    Thus, real drift is 5.5°/h and this is now entered under item 4. Working next step back, we know that the total apparent wander must be equal in magnitude and opposite in sign to the real wander. Therefore, the entry under total apparent wander is 5.5°/h increase.
    Complete the layout by inserting the apparent wander in 30°S (7.5°/h increase) and we have sufficient information to calculate convergency:

1. App. wander $30°S = 15° \sin 30° = 7.5°/h$ inc
2. Convergency = dep tan lat   = _____   ?
3. Total app. wander   = 5.5°/h inc

This gives the convergency value of $-2°/h$. A minus sign indicates direction WEST or a track of $270°(T)$, one answer. Departure is the aircraft's ground speed.

	No.	Log
Conv = dep tan lat		
$\therefore$ dep = $\dfrac{120^*}{\tan 30°}$	120	2.0792
	$\tan 30°$	$\bar{1}.7614$
= 207.9 kt		2.3178

$^*2°$ of convergency is converted to min of long to give $120'$.

13. A directional gyro has a drift of $8°/h$ (dec) when at $52°30'N$. On what true heading should the aircraft fly in latitude $30°N$ for the drift to be the greatest and what would be the drift rate if the TAS is 420 kt? Assume still air conditions.

Answer: Hdg 090(T); drift 7.6 decrease.

Solution

Maximum drift is always on the easterly track.
App. wander $52°30'N = 15° \sin 52°30' = 11.9°/h$ dec

Actual drift	= 8.0°/h dec
Real drift	= 3.9°/h inc
App. wander at $30°N = 15° \sin 30°$	= 7.5°/h dec
Conv = dep tan 30° = 7 × 0.5774	= +4.0°/h (east)
Total app. wander	= 11.5°/h dec
Real drift	3.9°/h inc
Total drift	= 7.6°/h dec

14. An aircraft's directional gyro is ground tested at $50°N$ and is found to have a drift of $8°/h$, the readings decreasing. A compass comparison during the pre take-off checks give the following readings: DI reading $300°$, H(T) $281°(T)$. The aircraft then flies a due south track for four hours at a ground speed of 180 kt. What is the DI reading at the end of this time if the true heading of the aircraft is $174°$?

Answer: $165\frac{1}{2}°$.

Solution
First establish the real wander due to the lat nut.

At $50°N$ app. wander = $15° \sin 50°$	= 11.5°/h dec
Actual wander	= 8.0°/h dec
$\therefore$ Real wander (lat nut)	= 3.5°/h inc

Next, establish the total wander per hour on mean lat basis.
GS = 180 kt $\equiv 3°$ of latitude;
$\therefore$ In 2 hours (half time) ch lat = $6°$

and mean lat = $50° - 6° = 44°N$.

App. wander $44°N = 15° \sin 44°$	= 10.4°/h dec
Convergency	= 0.0
Total apparent wander	= 10.4°/h dec

| Real wander (lat nut) | = 3.5°/h inc |
| Total wander | = 6.9°/h dec |

Lastly, establish the heading as follows:
On initial check immediately before take off

Gyro reading	= 300°
True heading	= 281°
Gyro reading is	19° too high
Total decrease	= 27.5° (6.9° x 4)
Gyro's present reading should be	8.5° too low

∴ When the true heading is 174°, the gyro heading is
174° − 8.5° = 165.5°.

Alternatively

Initially the gyro reading is 19° too high, that is, if the check was made on a true heading of 174°, the gyro reading would have been 174° + 19° = 193°. This reading has decreased by 27.5° due to the apparent and real wander. The reading now is
193° − 27.5° = 165.5°.

15. A direction indicator compensated at the equator gives zero drift while flying a rhumb line track of 270°(T) along the parallel of 54°23′N. What is the aircraft's ground speed?

Answer: 524.4 kt.

Solution
The aircraft is flying against the direction of the rotation of the earth. The change of longitude is such that it cancels out the effect of the earth's rotation. Therefore, the convergency due to the change of longitude is equal to the apparent wander due to the Earth's rotation.

App. wander at 54°23′N = 15° sin 54°23′ = 12.2°/h dec
Conv. = dep x tan 54°23′ also   = 12.2°/h
Resultant drift   = 0
Conv. = 12.2° = 732′ of longitude

$$dep = \frac{convergency}{\tan lat}$$

	No.	*Log.*
$= \dfrac{732'}{\tan 54°23'}$	732	2.8645
	tan 5423	0.1448
= 524.4 kt		2.7197

16: Artificial Horizon

Principle

An artificial horizon employs a vertical axis Earth gyro having freedom in all three planes and indicates the aircraft's attitude in pitch and roll. The gyro axis is maintained vertical with reference to the centre of the Earth so that a bar across and at 90° to the rotor axis indicates the horizon. In flight an aircraft rolls and pitches about the gyro axis which remains rigid and the indications are instantaneous.

Construction

The rotor of the gyro is encased in a sealed case which acts as inner gimbal. Air is let into the case under pressure, the pressure either created by a pressure pump or by creating suction inside the case. The suction required is 4″ of mercury. The rotor spins under air pressure at the rate of approximately 15 000 rpm. Having spun the rotor, the air escapes from the case through four exhaust ports in a pendulous unit mounted at the base of the gyro. Electrically driven gyros require a supply of 115V 3 phase 400 cycles AC and produce a spin rate of approximately 22 500 rpm.

The inner gimbal is mounted in the outer gimbal with its axis athwartships. The outer gimbal is mounted in the case with pivot points in the fore and aft axis of the aircraft. The inner gimbal having its movement about the athwartships axis controls the indications in pitch attitude. It has the freedom of movement of 55° either side of the central position. The outer gimbal controls the indications in the rolling plane (bank) and has freedom of about 90° from its central position. Later models of

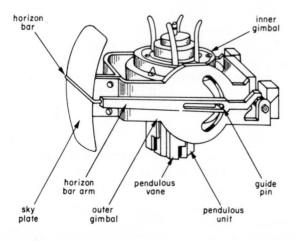

horizon bar

inner gimbal

horizon bar arm

pendulous vane

guide pin

sky plate

outer gimbal

pendulous unit

Fig. 16.1

electrical gyros have complete freedom in roll and 85° in pitch. With such gyros it is possible to do a complete loop without affecting the gyro rotor.

Operation

Any movement relative to the inner gimbal is transmitted to the horizon bar arm through a guide pin on the inner gimbal. The guide pin engages the horizon bar arm through a curved slot in the outer gimbal — fig. 16.1. During level flight the aircraft's vertical axis is parallel to the rotor axis and the guide pin is in the centre of the slot. Horizon bar arm is in the centre, and its extension across the face of the dial is in the centre of the dial behind the miniature aircraft. When the aircraft climbs or descends the rotor case (that is, the inner gimbal) remains rigid whereas the outer gimbal and the instrument case move with the aircraft. Due to the movement relative to the inner gimbal the guide pin gets displaced in the slot taking the horizon bar arm with it. Thus an indication of climb or descent results. Fig. 16.2(a) represents the relative positions of the miniature aircraft and the inner and outer gimbals in straight and

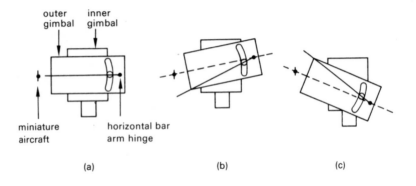

Fig. 16.2

level flight. Fig. 16.2(b) and (c) represent the relative positions in climb and descent respectively.

The bank indication is given by an index on the sky plate which is directly connected to the outer gimbal. The index reads against a scale printed on the glass face of the instrument. When an aircraft banks, the rotor, inner gimbal and outer gimbal remain rigid in level position and the instrument case together with the printed scale moves with the aircraft. Thus, the position of the index on the sky plate indicates the bank angle against the scale.

Gyro control

The erection mechanism of a pressure driven artificial horizon consists of a pendulous unit with its four exhaust ports, two in the fore-and-aft axis and two in the athwartships axis of the aircraft. Pivoted on top of the ports are four vanes, each covering half the port in straight and level flight. In this condition the air escapes equally from all four ports and the system is in equilibrium.

If the rotor axis departs from the vertical, the pivoted vanes will still remain in the true vertical with the result that one pair of ports will be out of balance (one port

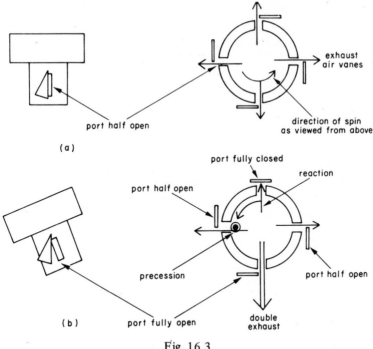

Fig. 16.3

will be more than half open; its opposite port similarly closed). In fig. 16.3(a) the rotor axis is shown as vertical, air escaping equally through all four ports. In fig. 16.3(b) the rotor axis is shown tilted. The vane on one port has fully uncovered the port and the vane directly opposite to it has fully covered up the port. Unbalanced airflow will take place as a result, setting up reaction in the direction of the closed port. This reaction force will precess through 90° in the direction of the spin and erect the gyro. The resulting rate of precession is deliberately kept low so that the rotor will not precess when the vanes are being thrown about in turbulent flight conditions.

Electrically driven rotors are controlled and kept in the vertical by means of a pair of torque motors and level switches. The torque motor correcting in the roll axis is mounted on the starboard side and the one correcting in the pitch axis at the rear of the instrument. Two level switches contain mercury liquid as a contact agent and the switches are placed at the base of the inner gimbal. When the gyro is level, the mercury is held in the centre of the trough, no contact is made at either of the two electrodes at the sides of the trough and the torque motor is switched off. When the gyro is not level, mercury rolls down to the lower sides and completes the circuit at one of the contacts. This energises the appropriate torque motor which applies a torque to the rotor in the correct direction and erects the gyro (fig. 16.4).

Errors

Unlike the DGI an artificial horizon suffers from both acceleration and turning errors. In the following discussion the rotor of the artificial horizon is assumed to turn anti-clockwise as viewed from above.

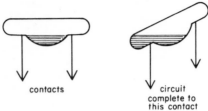

contacts   circuit
                            complete to
                            this contact

Fig. 16.4

Acceleration Errors. These are also known as 'take-off' errors since they are most noticeable during this phase of the flight. There are two elements which introduce the errors: the pendulous unit and the vanes.

The pendulous unit makes the rotor bottom heavy. Thus when an aircraft accelerates, a force due to the unit's intertia is felt at the bottom, acting towards the pilot. This force precesses through 90° in an anti-clockwise direction and lifts up the right hand side of the outer gimbal. The skyplate attached to the outer gimbal rotates anticlockwise, the bank index indicating a false starboard turn.

Also during acceleration both port and starboard side vanes are thrown back with the result that the starboard side port opens up fully and the port side closes down fully. The reaction occurs on the port side, precesses through 90°, and lifts up the inner gimbal from the point nearest to the pilot to indicate a false climb.

Turning Errors. During a turn the fore and aft vanes will be displaced due to the centrifugal force in a direction away from the centre of the turn. Thus, one port will be open and its opposite one closed. The reaction will be set up in the fore and aft axis of the aircraft which will precess through 90° to lift up the outer gimbal at the port or starboard side. This results in indication of false bank. The sense of the indication depends on the direction of the turn. This particular error is also known as erection error.

Centrifugal force also acts on the pendulous unit, the force acting from starboard to port or vice versa, depending on the direction of the turn. This force affects the inner gimbal, giving a false indication of climb or descent. This error is also known as pendulosity error.

The combined effect of the two is to displace the gyro rotor in two planes. If a turn is made through 360° the error reaches maximum at 180° and then starts reducing until the turn is complete when the error will have reduced to zero. In modern gyroscopes, the rotor axis is displaced from the true vertical to counteract these errors. But the correction is only valid for a given rate of turn and a given speed. For example, a pressure driven gyro, offset $2\frac{1}{2}°$ forward and $1\frac{3}{4}°$ to port corrects for a rate one turn at 190 kt. The tilt does not affect straight and level indications as the scales are similarly offset.

Operational limits

If the gyro limits are exceeded the gyro will topple. Therefore, where the facility is available, the gyro should be caged before entering severe manoeuvres. An electrically driven gyro with 85° freedom in pitch may be taken through a loop without affecting the rotor. As the loop angle progresses between 80° and 100°, the horizon bar smartly travels across the face of the instrument from the bottom to the top. The skyplate bank index will similarly appear at the top. To the pilot who is inverted the indications will look correct. (What is the normal top is the bottom when looking at it upside down if you see what we mean.) A similar change in the opposite direction takes place when the loop is going through the 260°–280° zone.

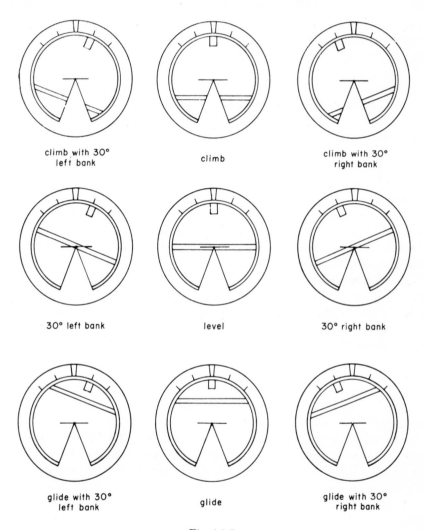

climb with 30°
left bank

climb

climb with 30°
right bank

30° left bank

level

30° right bank

glide with 30°
left bank

glide

glide with 30°
right bank

Fig. 16.5

17: Turn and Slip (Balance) Indicator

These two instruments are generally combined, with two pointers giving two types of information read in conjunction with each other.

TURN INDICATOR
This part of the instrument indicates the rate of turn and utilises the principle of gyro precession to do so.

Construction
The instrument employs a horizontal axis gyro, the rotor being mounted in a horizontal ring in the athwartships axis. This ring itself is mounted in the fore-and-aft axis of the aircraft in the instrument case. Thus, the gyro has freedom of movement about one plane only, that is, about the fore-and-aft axis. The rotor is driven by an air jet which impinges upon the buckets on the rotor rim and spins it at an approximate rate of 9 000 rpm. This speed is considerably lower than DGI or Artificial Horizon speed, for the reasons that the Turn indicator utilises the principle of precession for its operation and that the instrument incorporates two springs which hold the gyro axis horizontal in level flight. The rotor spins away from the pilot.

Operation
During straight and level flight the springs hold the gyro axis horizontal preventing unwanted precession, and the pointer attached to the vertical rotor indicates central or zero position on the instrument against the printed scale (fig. 17.1).
    As the aircraft enters into a turn, say a turn to the left, the gyro axis being rigid, opposes the turn and a force is experienced on the axis, as shown in fig. 17.2(a). On the left hand axis the force is coming out of the paper. Let us call this force 1. This

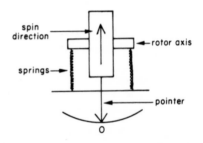

Fig. 17.1

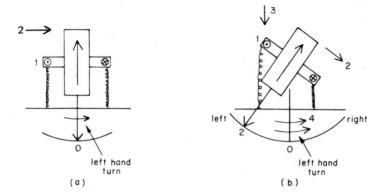

Fig. 17.2

force will precess through 90° and act at the top of the rotor, force 2, causing the rotor to tilt. This tilt is called primary precession. If no springs were attached to the rotor axis, the rotor would continue to tilt until it is spinning in the horizontal plane, with its axis vertical. This attitude would give no indication of the rate of the turn. However, having the springs, as the rotor starts to tilt, one spring stretches and the other contracts. In a left hand turn, as shown in fig. 17.2(b) the left hand spring is stretched. This produces a pulling down force on the left hand axis or a pushing up force on the right hand axis — force 3. This vertical force on the axis will precess the rotor in the horizontal plane in the direction shown — force 4. This is called secondary precession. If you notice, force 4 acts in the same direction as the direction of the turn. As the rate of turn is established, force 1 becomes a constant value. When force 4 reaches the value of force 1, we have a situation where two forces of equal value are acting for opposite purposes (force 1 is due to rigidity of the axis; force 4 is precession force). At this stage, the gyro cannot tilt any further. Whatever tilt has duly occurred is entirely due to force 1, which in its turn, is due to the rate of turn. Therefore, the gyro tilt together with the pointer displacement represent the rate of turn.

If the instrument case is not airtight, air will be drawn in from leaking points due to suction inside the case and loss of efficiency will result. If the rotor speed is less than the rated speed, the pointer will indicate a lesser rate of turn; similarly, if the speed is too high, it will indicate a higher rate of turn.

The spring tension is adjusted for a given rate of turn, usually rate one, and at all other rates the indication will be progressively in error.

Movement in the looping plane theoretically will have no effect on the gyro since the aircraft loops about the gyro axis and not against it as it does when it enters into a turn.

SLIP INDICATOR

This part of the instrument is entirely mechanical and depends on the forces acting on a pendulous weight for its indications. In straight and level flight the pendulous weight is acted upon by the force of gravity which keeps it in the true vertical. The pointer attached to the system indicates the central position, that is, no slip or skid.

During a turn the weight is acted upon by two forces: gravitational force acting downwards and centrifugal force acting away from the centre of the turn. The pendulum takes up the position which is the resultant of the two.

If the turn is a balanced turn, the two vectors must be of such dimensions as to shift the weight from the true vertical to the aircraft's vertical. This is so, because in a balanced turn weight opposes lift, and the lift acts at 90° to the wings, that is, in the aircraft's vertical axis. Therefore, weight must lie in the vertical axis (fig. 17.3).

In fig. 17.4, aircraft is shown in an RH turn with insufficient bank. The aircraft skids outwards from the centre of the turn. The pointer is displaced in the opposite direction to the direction of the turn.

In fig. 17.5, aircraft is in an RH turn, having an excessive bank. The aircraft slips inwards, the pointer is displaced to the right of the zero, that is, in the same direction as the direction of the turn.

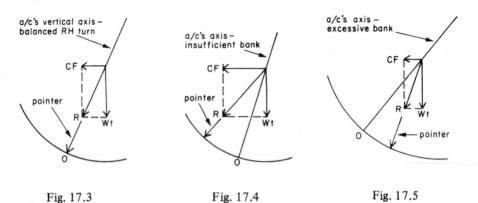

Fig. 17.3   Fig. 17.4   Fig. 17.5

Some models employ a ball in the tube arrangement in place of a pendulum. The ball itself has weight (it is not a bubble) and thus, it is affected by the aircraft's relevant manoeuvres in exactly the same way as the pendulum weight and the resulting indications are alike, except that the ball itself indicates slip or skid instead of a pointer.

From the above, the rule of interpretation is that if the rate of turn pointer and slip pointer (or the ball) are displaced in the same direction, the aircraft is slipping inwards towards the centre of the turn. If the two are displaced in the opposite direction, the aircraft is skidding outward. If the turn is indicated but the slip pointer is at zero position, it is a balanced turn.

Serviceability check

While standing on level ground both pointers should indicate zero position. While taxying, check indications by a slight turn. The turn should be indicated in the correct direction and slip pointer should indicate a skid (insufficient bank!).

Electrical and pressure driven gyros

Electrically driven gyros have some distinct advantages over air driven gyros, the main advantage being that it is possible to run electrical rotors at higher speeds. As we saw

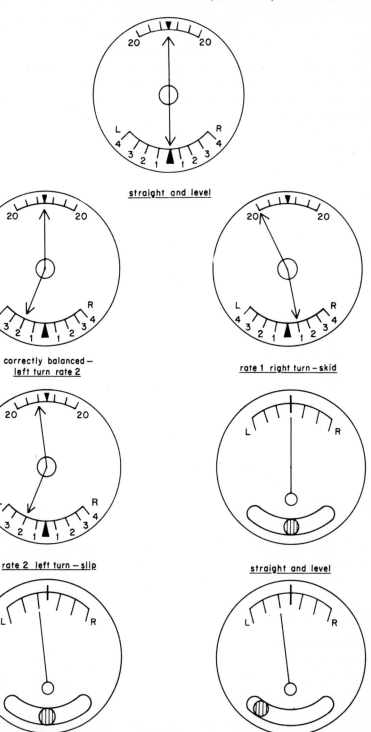

straight and level

correctly balanced —
left turn rate 2

rate 1 right turn — skid

rate 2 left turn — slip

straight and level

rate 1 left turn — balanced

rate 1 left turn — slip

Fig. 17.6

earlier, higher speeds impart greater rigidity and therefore greater accuracy where rigidity is utilised as the operating principle of the instrument. From the designing angle, electrical gyros yield neater designs since air passage tubing and filtration devices are no longer required. This permits gimbal designs giving greater operational limits. Further, since air no longer enters the rotor unit, there is freedom from filtration troubles and moisture corrosion. Finally electrical rotors are not affected by rarer atmosphere at high altitudes.

18: Inertial Navigation

This is a technique for determining a vehicle's position and velocity by measuring its acceleration with respect to a known set of axes, and processing that acceleration in a computer. This self-contained navigation is now in reasonably common use by airlines whose routes justify it and who can afford it, especially after sacking their navigators. It has no need for external references, has nothing to do with magnetism, no plotting is required, it is unaffected by weather, all corrections for movement over the earth's surface and the movement of the earth itself are integral, and position in latitude and longitude, distance flown and other navigational information are presented before your very eyes digitally. And providing the vital preflight setting-up is correctly done, there is no possibility of human error.

The term *acceleration* means rather more than just increasing speed as it is considered in the loose phraseology of daily life; it is changes in velocity, and velocity is speed measured in a definite direction along a straight line; and acceleration of a body is directly proportional to the sum of the forces acting on it. *Inertia* is the property of matter by which it continues in its existing state of rest or uniform motion in a straight line unless that state is changed by external force. An *accelerometer* is the device that measures the force required to accelerate a mass, i.e. it measures the acceleration of the vehicle containing the accelerometer. The motion of an aircraft in inertial space can be determined therefore from information contained within the aircraft itself by measuring the accelerations in the north/south and east/west directions, producing velocities and displacements which can then be processed to give position and ground speed, to name but a few.

An accelerometer is at rock bottom a pendulum which moves off the vertical when the aircraft accelerates: the amount of movement from the state of rest can be measured, amplified, and current from the amplifier is returned via a torquer to bring the pendulum to the null. The amount of current required is a primary indication of the acceleration experienced.

The signal also goes to an integrator, where it is multiplied by time: in other words, the acceleration is measured in feet per second squared, and the integrator turns it into feet per second: it is simple to transfer this via a second integrator into miles for the time flown. Thus a single accelerometer has found how far the aircraft has gone in a definite direction for a definite time. If two accelerometers are used, one in the north/south axis, the other in the east/west, then from a definite start point in latitude and longitude, the new position of the aircraft can be shown.

In its simplest terms, the system has three accelerometers in an aircraft, each held irrevocably in its chosen reference axis, north/south, east/west and local vertical; each measures the acceleration along its axis, and the integration of this acceleration with respect to time will provide the velocity along this axis, though initial con-

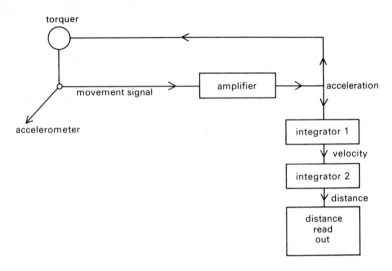

Fig. 18.1

ditions must be known. A computer converts the time integrals along each axis into terms of distance travelled, and still further into latitude and longitude.

The system
The platform consists of a gyro-stabilised cluster of accelerometers whose outputs are fed to a computer. The platform isolates the accelerometers from angular rotations of the aircraft, and maintains them in a fixed orientation relative to the earth, usually with two of the input axes being locally level. The structure, gimbal-mounted, on which the gyros (a pair of two-degrees-of-freedom) and three accelerometers are mounted, is called the *stable element*: the gimbals allow the aircraft to rotate without disturbing the stable element by detecting any movement from the level, and amplifying the signal to a gimbal drive motor which restores the gimbal assembly, i.e. the platform, level. Keeping the accelerometers exactly in the chosen reference axes, with an immediate reaction to the smallest deviation therefrom, is not regarded as much of a problem, using accurate gyros and high-performance servo mechanisms. The gyros act as error detectors to sense inadvertent rotations of the stable element, and apply corrections to gyro torque motors to cause the appropriate precession.

The computer calculates the aircraft's position and velocity from the outputs of the two horizontal accelerometers; it also calculates the gyro-precession signals which maintain the stable element in the desired orientation relative to the earth, since the earth itself is rotating and the aircraft is rotating about the earth, so the platform must rotate at these combined rates to stay aligned with the chosen earth's axes; it also calculates accelerations other than those caused by changes in the aircraft's motion relative to the earth.

In each case, the computer initiates the required compensation. When the computer is turned on, it must be set up so that it knows the initial position of the aircraft; the stable platform must have the correct initial orientation relative to the earth; the platform is typically aligned in such a way that its accelerometer input axes are horizontal, often with one of them pointing north. A vertical accelerometer is sometimes added to speed up the indication of altitude on the dial as measured by the barometric altimeter or air-data computer.

The calculation of velocity from the outputs of the horizontal accelerometers

This involves some pretty terrifying mathematics, which you might enjoy but won't get here. It is necessary to derive the velocity of the aircraft relative to the ground, a ground which is part of a rotating earth. The first co-ordinate frame is centred at the earth's centre, projecting through the essential reference points of the equator, Greenwich and the pole, and considered to be motionless with regard to the stars. The second co-ordinate frame is similarly centred, but fixed to the earth. The axes of both will be aligned only at the instant of alignment, for the equatorial axis of the latter will move at $15.04°$ per hour relative to the former, and only the polar axis will remain near-enough coincident. The third set is the actual co-ordinates of the aircraft itself at a given moment, one axis east/west, another north/south, another vertical, all radiating as it were from the aircraft itself.

Furthermore, the platform has its three co-ordinates with respect to the input axes of the accelerometers, at right angles to each other, set north-pointing or in the aircraft's axes lines. Thence, with the time factor since alignment of the system, the output of the accelerometers is solved in the computer to give a speed over the ground from a complex working of:

(a) the initial angular velocity of the platform

(b) the components of the platform axes proportional to the accelerometer outputs

(c) the angular velocity of the earth in inertial space

(d) the elimination of unwanted accelerations of gravity (not constant), of coriolis and centifrugal forces (the tendency to move outwards due to following a curved path in space in order to maintain a great circle path on a turning earth): these will be picked up by the accelerometers, which cannot differentiate, but are processed and rejected by the computer.

A plain diagram will show the gyro-processing scheme to keep the platform level with the earth's surface as the aircraft moves over the rotating earth. Briefly, the sum of the two compensations reaches the gyro torquer and the amount of tilt thus

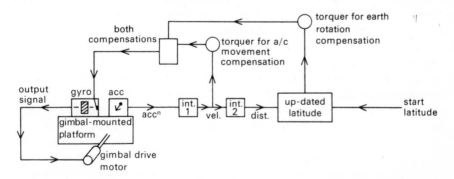

Fig. 18.2

caused is measured and amplified: this current, which is in proportion to the amount of tilt, is used to drive the gimbal motor which causes the gimbal to tilt and regain the correct horizontal at the aircraft's new position on the earth's surface. A further torquer is placed between the integrators to correct for coriolis, ellipticity of the earth and other sundry compensations consequent of flying over an imperfect rotating sphere.

In providing information about altitude, the vertical accelerometer does not perform its function with quite the facility of the horizontals. A small error in the early stage, whether of initial altitude, of altitude rate, or of the mechanisation of the vertical gravity component, will increase considerably with time, rendering the indications useless even after a few minutes. A baro-inertial altimeter is used therefore, a combination of the two, which utilises the basic stability of the barometric altimeter, but avoids the time-lag from measuring altitude to its dial appearance, utilising the accelerometer to do this.

The calculation of position from the velocities

From the sets of co-ordinates already mentioned and the measurement of ground speed, it becomes a simple matter for the computer to solve the position of the aircraft in terms of latitude and longitude continuously in flight. The resolved components of velocity provide the essential ingredients of position finding, since they are speeds in a given direction, and the direction as a northing or easting element is as intrinsic as the speed, all of which are instantly readable in the normal accepted figures.

Alignment

It is vital that the platform is accurately levelled and aligned in azimuth before take-off and so maintained during flight.

Platform orientation co-ordinates

(a) levelling until the output from the accelerometer is zero: the short-cut electronics involved eventually move the gimbals to move the accelerometer pendulum to zero, computed therefore to zero velocity.

(b) Lining the system up to true north, known as *gyro-compassing*. Again, with automatic compensation, this cannot be hurried. Compensation for the earth's rotation at that latitude is fed *in full* to the east/west gyro, the computer as it were assuming the north/south is accurately aligned: if there is a misalignment, however, the gyros are receiving a wrong compensation and the platform will tilt as the earth rotates: this tilt is automatically corrected, the platform points to true north, the gyros will receive the correct compensation, keeping the correct alignment in flight. Gyro-compassing need not always refer to true north. The north system is useless in high latitudes and over the pole, so an arbitrary angle can be set. The alignment procedure is much the same; the angle is primarily assumed to be zero, all the earth rotation compensation is picked up by the east/west gyro, and of course the platform will tilt, as the assumption is incorrect. This time the accelerometers actuate a torque which is split between the two gyros, eventually to determine the right combination to compensate for the earth's rotation for the particular angle required: the ratio to the gyros is then applied to compute the initial angle. This offset is coupled with the accelerations sensed in flight, and the computer knows the amount of offset.

Pre-departure
1. Align — this in fact continues while pre-departure procedures are being carried out.
2. Set STBY, stand by.
The selector has the following positions:

TEST
DSF TK STS   desired track angle.
WIND
DIS/TIME   distance to go/time to go.
WPT   waypoint in lat/long.
POS   aircraft's position in lat and long.
XTK/TKE   cross track distance from that required/difference between actual and desired track angle.
HDG/DA   aircraft heading/angular difference between HDG and TMG.
TK/GS   TMG/GS.

3. Test the CDU (control display unit) to ensure all points on it light up.
4. Enter the present latitude and longitude.
5. Insert waypoint latitudes/longitudes; this is best done while:
6. setting mode selector to 'align' to check alignment progress: and the display selector to DSR/TK/STS. On the right display, the number will have decreased from 90 (it's started) down to 02 (all finished, all well), and the READY NAV light will come on. Switch to NAV; and ready to go.

Of course, there are a number of refinements, and a number of differences with various types of presentation: the substance is scarcely variable, though. The only input required is the start latitude/longitude, nothing else: should this be entered sizeably in error, the compensations for earth's rotation and aircraft movement on a great circle track, to name but a few, will be incorrect and you might as well get your sextant out: you can't get a correct updated position from an incorrectly inserted initial position on which the gyro-torquing was based.

Waypoints
These are positions along the route which can be inserted (in lat/long) pre-departure; up to 9 of them, and stored in the computer. WPT is selected, and the number of the waypoint corresponds to that shown on the display unit. Once NAV has been selected, the track and distance can be displayed between any two waypoints by switching to RMT (remote) and thus providing a check against the flight plan of correct insertion of waypoint co-ordinates.

To sum up
1. The inertial system measures and integrates the accelerations experienced in flight by two accelerometers at right angles to each other.
2. These accelerations are computed from velocity and time to latitude and longitude of the aircraft's position, starting from the aircraft's start position.
3. The accelerometers are kept level on a stable platform by a system of gimbals and

gimbal torques: thus only motions tangential to the earth's surface are measured.
4. Correction for the rotating earth, the aircraft's movement over a spherical surface, the earth not being a perfect sphere, coriolis, centrifugal and centripetal acceleration are compensated for automatically via a computer.
5. All track, and return to track, requirements can be computed, displayed or fed to the autopilot.

Errors

In-built errors of the inertial system are very few, and 1 nm/h is bandied around as a never-exceeded error with some confidence. There is, though, a slight increase on this as a trip stretches out. An error is expressed as contained within the radius of a circle centred on the ramp position; more pungently and fashionably called the 'radial' error. To calculate this from the lat and long of the ramp position, vis-à-vis the INS position display is simply the solution of a right-angled triangle: distance on the latitude scale is okay, but you'll use the departure formula at mean latitude to bring ch long to distance. We're not piling it on too strong, are we?

Advantages of the inertial system

1. Indications of position and velocity are instantaneous and continuous.
2. Utterly self-contained, with no need of ground stations or whatever.
3. Navigation information is obtainable at all latitudes and in all weathers.
4. Navigation information is substantially independent of aircraft manoeuvres.
5. Any inaccuracies may be considered minor as far as civil air transport is concerned.
6. Apart from the overriding necessity for accuracy in pre-flight requirements, there is no possibility of human error.

Disadvantages

1. Position and velocity information does degrade with time: and again, this is true stationary or airborne.
2. Equipment is far from cheap, and is difficult to maintain and service.
3. Initial alignment is simple enough in moderate latitudes when stationary, but difficult above 75° lat and in flight.

MAGNETISM AND COMPASSES

19: Magnetism: General and Terrestrial

Magnetism: general and terrestrial

This is an interesting study, and of great importance: at last the divorce from the boatmen has been accomplished, though the decree nisi was long enough, Heaven knows. The student pilot need no longer learn about Kelvin's balls, but about air-borne compasses which are now highly accurate and trouble free. Yet a simple magnetic compass must always be aboard just in case. So off we go from the fundamentals of magnetism.

A magnet has the following properties:

1. It attracts iron filings, and more strongly at the poles. The attraction can be exerted at a distance, the effect decreasing with increasing distance.
2. When a magnet is freely suspended, one end tends to point in a Northerly direction and is called the North-seeking or RED Pole.
3. With two magnets near each other, the red pole will tend to attract the blue pole of the other, while like poles repel each other.
4. The amount of possible magnetism is limited by the mass of magnetic material in the magnet. When a magnet can no longer by any means be made more powerful, it is said to be saturated.
5. No magnet can exist with only one pole: even if broken into pieces, each piece would become a complete magnet with poles of equal strength.

The Molecular Theory supports this: any old piece of iron consists of molecules, each of which is a magnet but exerting its magnetism haphazardly giving no resultant magnetism to the iron bar. On magnetisation, however, each molecule is lined up so that all of them exert their magnetism in the same direction; with every one lined up, the bar is now a magnet, and magnetically saturated, at its maximum magnetic strength.

The sphere of influence of a magnet is called its magnetic field, composed of its magnetic lines of force. These are the lines the direction of each being the path in which an isolated red pole free to move would travel. They are quite definite, and never cross, since they depend on their position with regard to the attracting blue pole, the repelling red pole. Thus, a magnet lying across the magnetic field of another magnet would tend to take up the direction of the line of force running through it with the opposite poles nearer to each other.

Hard and soft iron

Any metal which can be magnetised at all comes under one of these headings: soft iron can be easily magnetised and will just as readily lose its magnetism when the magnetising influence is removed. Hard iron is difficult to magnetise, but once done tends to remain permanently so. Hard iron magnetism is called 'permanent' while soft iron is called 'temporary' or 'induced'.

The Coercive Force is the power which hard iron has of resisting magnetisation, or if already magnetised, of resisting de-magnetisation.

Terrestrial magnetism

The Earth itself is a magnet, with its own magnetic field, with a blue pole in the vicinity of the True North Pole: thus a freely suspended compass needle will line itself up with its red pole pointing to the Earth's Magnetic North Pole.

The Magnetic Poles are areas on the Earth's surface where a freely suspended compass needle, influenced only by the Earth's magnetic field will stand vertical.

The Angle of Dip in this case is 90°; the Angle of Dip is defined as the angle between the horizontal and a freely suspended compass needle influenced only by the Earth's magnetic field. Thus the Magnetic Equator is an imaginary line on the Earth's surface joining all points where the Angle of Dip is Nil, and can be said to be the dividing line between the Earth's blue and red polarity.

The Earth's line of total force at any particular place is the line of force in which a freely suspended compass needle lies in the Earth's magnetic field when influenced only by the Earth's magnetic field. It follows that only at the Magnetic Equator will the needle be horizontal, only at the Pole will it be vertical. The Directive Force is that component of the Earth's total force which acts in a horizontal plane, known as H; likewise the Vertical force in the vertical plane, known as Z. The Magnetic Meridian is simply the direction in a horizontal plane of the freely suspended compass needle influenced only by the Earth's magnetic field, i.e. the direction of the directive force. Thus, we have a triangle of forces depending on the position of the compass needle in the Earth's magnetic field which solves the actual horizontal direction and angle from the horizontal which it will actually take up along the line of the Earth's total force.

A change of magnetic latitude will thus affect the freely suspended needle: the nearer to the Pole, the greater Z, and the less H: conversely, near to the magnetic equator, Z is small, H is high. The triangle thus:

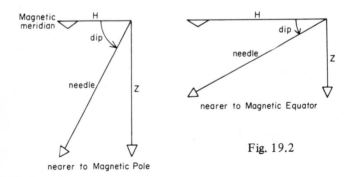

nearer to Magnetic Pole

Fig. 19.1

nearer to Magnetic Equator

Fig. 19.2

An examination of figs. 19.1 and 19.2 will reveal the following relationships:

(a) Tan dip $= \dfrac{Z}{H}$

(b) $Z = H$ tan dip

(c) $H = \dfrac{Z}{\text{tan dip}}$

(d) Sin dip $= \dfrac{Z}{T}$, where T is the total force, and

(e) $T = \dfrac{Z}{\text{sin dip}}$

Now we will introduce a few problems which require use of the above formulae.

1.   At a place A, the dip is 63° and the value of H is 0.18 units. What is the value of Z and the total force T at this place?

$$Z = H \tan 63°$$
$$= 0.18 \times 1.9626$$
$$= 0.3533$$
$$T = \frac{Z}{\sin 63°}$$
$$= \frac{0.3533}{0.8910}$$
$$= 0.3965$$

2.   At a place X, the value of H is 0.22 and the value of Z is 0.44 units. What is the angle of dip at this place?

$$\text{Tan dip} = \frac{Z}{H}$$
$$= \frac{0.44}{0.22}$$
$$= 2.000$$

∴   Angle of dip = 63°26′

3.   Given, dip = 60°, Z = 0.27; find the values of H and T.

$$H = \frac{Z}{\tan 60°}$$
$$= \frac{0.27}{1.7321}$$
$$= 0.1559$$
$$T = \frac{0.27}{\sin 60°}$$
$$= \frac{0.27}{0.8660}$$
$$= 0.3118$$

Now you try this — you will need to manipulate the formula a bit.
Given, dip = 61°, H = 0.156, what is the value of T? Check that you are getting the answer 0.322.

Changes in the earth's magnetic field

The earth's magnetic field at any point is defined from the knowledge of its three characteristics, that is, variation, intensity and angle of dip. Such a field is very irregular and seems to be forever changing. There are long-term changes, periodic changes, local anomalies and a magnetic storm may play havoc.

Secular change. This is the long-term change and reasonably predictable. At one time it was thought that this occurred due to the rotation of the magnetic pole round the geographical pole. Unexpected changes have been recorded at different places and the overall results do not bear out this theory entirely. Since these changes are more of a localised nature, they might be caused by changes in the currents flowing deep within the earth's interior in a particular region.

Of these changes, as aviators we are primarily interested in the changes in variation. The predicted rate of annual change is given on the maps, either by a statement in words, e.g. 'annual change 7'E, or by an arrowed line across the isogonal with a numeral on it. The arrow head indicates the direction of the change and the numeral gives the number of nautical miles movement per year.

Periodic changes. Irregular changes occur daily, annually and every 11 years.

Daily changes appear to be caused by the electric currents flowing in the ionosphere. These currents are the result of the 'atmospheric tides', the sun's heating action providing the energy. The tidal currents in the ionosphere then modify the earth's field. The changes swing between 4' on a quiet day to 12' on a sun-spot day, the maxima occurring just before sunrise and at noon.

The annual change takes place in the form of a cycle varying between 2'E to 2'W, the maxima occurring around 21st March and 21st September. The changes in the southern hemisphere are opposite in sense and it is thought that the earth's motion round the sun is concerned with this type of disturbance.

The 11-year changes, like the daily changes are positively associated with the 11 year sun-spot cycle. When the activity is at a maximum, the disturbances are also at a maximum.

Local anomalies. Strictly speaking no two places may be expected to have exactly the same field. This is due to the varying rock formation both on the surface and the interior. However, these differences are by no means remarkable and do not draw the attention of the aviator. On the other hand there are places, blessed with an abundance of ferromagnetic substances, which clearly distinguish themselves from the neighbouring regions, causing deviation limited to the locality and only near the surface. Greenland for instance is a tidy magnet in its own right, and there is a small area in mid-Atlantic, a few miles in diameter too. A place near Port Snettisham in Alaska causes a difference of 60° in the compass direction on the surface: it reduces rapidly with height (15° at 3 000 ft).

Magnetic storms. Associated with sun-spot activity magnetic storms may last from a few hours to several days. The intensity varies from 'small' to 'very great' and may envelop the whole earth in a matter of minutes. The effect on the aircraft's compass varies with intensity, but both variation and the horizontal directive force are modified. Generally the effect is more severe in higher latitudes; a deviation of 20° by the needle when flying near Iceland in a magnetic storm is on record.

The regular changes can be allowed for when the isogonals are plotted for an area — those lines joining places of equal magnetic variation, and the amount of annual change of variation forecast: in the U.K., the total change is about 7'E, and is hardly likely to be of significance to the pilot until he uses a chart more than 5 years old. Remember the definition of variation? The angular difference between the true and magnetic meridian at a point measured in degrees East or West of True. A line joining places of nil variation is the agonic line. And isoclinals are lines joining all places of equal magnetic dip.

To redefine deviation, it is the angular difference between the magnetic meridian and the direction taken up by a particular compass needle, measured E or W of the magnetic meridian. Aircraft are full of disturbing magnetic influences, and considerable thought is given by constructors and designers to reduce them. Fig. 19.3 should refresh your memory.

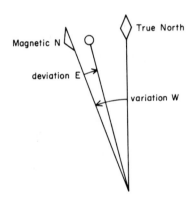

Fig. 19.3

Summary of the definitions

Isogonal.   A line on the surface of the earth along which all points have the same variation.

These lines generally face the magnetic pole but it is not always the case.

Agonic line.   A line on the surface of the earth along which all points have nil variation.

Variation.   This is the angle between the horizontal direction taken up by a freely suspended magnetic needle under the influence of the earth's magnetic field alone, and the direction of true north. It is named east or west according to whether the north-seeking end of the needle lies to the east or west of the true meridian.

Magnetic equator or aclinic line.   The line along which all points have zero dip.

It runs close to the equator: to the south of the equator in South America and to the north of it in Africa and the Orient.

Isoclinal.   It is a line along which all points have equal dip.

20: Aircraft Magnetism 1

An aircraft contains both hard and soft iron giving rise to permanent and temporary magnetism.

Permanent magnetism in aircraft arises chiefly from hammering whilst under construction and the effect of the Earth's line of total force running through it during building. The molecules in the hard iron tend to line up in the direction of the Earth's line of total force giving red polarity at the north-pointing end. The nature of this permanent magnetism depends on:

1. Magnetic heading during construction.
2. The Angle of Dip at the place of construction.
3. Amount of coercive force of the metals used.
4. And of course the amount of hammering, battering and riveting.

Some permanent magnetism can also be set up by the introduction of electro-magnetic material, radio and radar equipment, electric currents. A demagnetising effort called 'de-gaussing' is made before an a/c is sent from the workshop, but there is always some residual magnetism.

This permanent magnetism and its effect on the magnetic compass can be measured per aircraft, and a suitable correction card placed nearby. The 'Induced' or temporary magnetism in soft iron components of the structure (brought about by the Earth's field, giving these components a changing magnetic value depending on the variable strength of H and Z as the aircraft goes about its business) is not so readily taken care of.

In order to analyse the effect of permanent magnetism, we imagine it to be due to 3 components, acting respectively in the fore and aft, athwartships and vertical line of the a/c.

P is the parameter (or component) acting in the fore and aft line of the compass named + when the blue pole is forward of the compass.

Q is the parameter acting in the athwartships line of the compass, named + when the blue pole is starboard of the compass.

P and Q are horizontal hard iron, then, in the horizontal line of the compass needle itself.

R is the parameter acting in the vertical line of the compass, named + when the blue pole is beneath it. Its effect on the compass in straight and level flight may be taken as negligible.

Analysis of + P

Imagine the a/c on 8 headings, with a blue pole in the nose, red pole abaft the compass, pardon the expression.

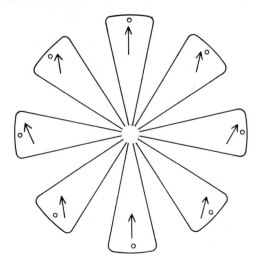

Fig. 20.1

Drawing up a table of deviations, and a graph of them:

Hdg(M)	Deviation
N	No deviation; directive force increased
NE	Easterly deviation
E	Max Easterly deviation
SE	Easterly deviation
S	No deviation; directive force decreased
SW	Westerly deviation
W	Max Westerly deviation
NW	Westerly deviation

Fig. 20.2

and we have a sine curve; on the quadrantal Headings Magnetic the deviation will therefore be .7P. A parameter of negative value, −P, would give a Westerly deviation on Easterly Headings Magnetic. We can summarise then by saying that the deviation due to P is proportional to the sine of the aircraft's Heading Magnetic, or

Deviation = P sin Hdg(M)

and P is positive or negative depending on Easterly or Westerly deviation with the aircraft on Headings Magnetic between North and East.

Analysis of −Q

The red pole is to starboard of the compass. You draw it out and agree the following graph:

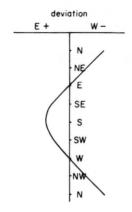

Fig. 20.3

This is a cosine curve, and the sign of Q is determined by the sign of the deviation obtained on aircraft Hdgs(M) from North to East.

Q is proportional to cos Hdg(M), therefore:

Deviation = Q cos Hdg(M)

and again the deviation on the quadrantals is .7 of the parameter, in this case .7Q.

P and Q are horizontal hard iron, permanent: we can conveniently co-relate with them certain soft iron components which act in the same manner on the compass; these induced effects are dependent on the strength of the Earth's magnetic field at the place, and of course on the coercive force of the soft iron itself.

These parameters are 'c' and 'f' ('little c and little f'):

'c', a pair of vertical soft iron rods one before the compass, one behind, each with a pole in the horizontal plane of the compass needle. In the Northern Hemisphere, the bottom pole would be induced with a red magnetism, and we assume the forward vertical rod to have its blue pole level with the needle, the rear rod its red pole, then we have effective poles acting on the compass like +P.

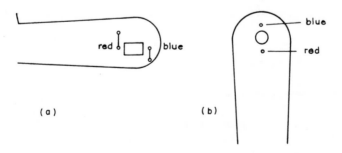

(a)   (b)

Fig. 20.4

Similarly for 'f', this time athwart the compass.

Vertical soft iron will not change its polarity with change of heading; thus, we consider P + c and Q + f at the same time for this analysis.

Proceeding to find a method of correction of the deviations obtained to render the compass as extensively serviceable as possible, the value of the deviations caused by P + c is resolved into Coefficient B, in degrees of deviation.

Coefficient B = P + c

which has maximum deviation on Easterly and Westerly Headings: thus, the Coefficient can be rendered as a numerical number of degrees by

$$\text{Coefficient B} = \frac{\text{Deviation on E} - \text{Deviation on W}}{2}$$

for the deviation on Westerly Heading is expected to be of opposite sign to that on Easterly, and we wish to remove the mean maximum deviation. The sign of the deviation is ready for correction on East Magnetic. Similarly,

Coefficient C = Q + f

and is calculated by:

$$\text{Coefficient C} = \frac{\text{Deviation on North} - \text{Deviation on South}}{2}$$

The value is given a sign ready for correction on a North Heading (M). Thus, while not expecting complete perfection of compass reading, we can eradicate the effect of fore/aft and athwartship hard iron magnetism, and vertical soft iron magnetism present.

Once found, coefficients B and C can be to some extent neutralised by setting up a local magnetic field: this correcting device is called the micro-adjuster, and is placed directly under the compass as an integral part of the instrument. Two pairs of magnets, one pair fore and aft, the other pair athwartships; the magnets of each pair can be opened like scissors from the null position to form effective poles at right angles to the original line of the magnets. A key is inserted to operate either pair of corrector magnets, and the rule is to use the key hole which lies at right angles to the compass needle.

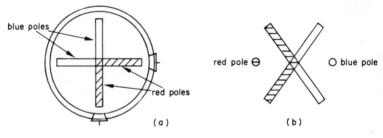

Fig. 20.5

Fig. 20.5(b) illustrates the opening of the fore and aft pair to give an effective blue pole to starboard, which would correct a negative Coefficient C (Q + f).

Coefficient A

This is caused by deviations which are the same on all headings, and nearly always is due to incorrect mounting of the lubber line. The rear of the compass bracket is screwed in against a span of 10° either side of centre, and the mounting of 0° may not be in the fore and aft axis of the aircraft: this would lead to a constant deviation on all headings: a Heading of 000(M) giving a compass reading of 010(C) entirely due to incorrect mounting would give a deviation of −10, which would be repeated on all headings, since the lubber line is displaced to starboard. This is called Apparent A, due to mechanical error. Real A, due to Horizontal soft iron, giving a constant graph of deviation too, is fortunately not part of our study.

Coefficient A is deduced by the deviations on 8 headings, adding them algebraically, and dividing by 8.

Thus, the deviation on any Heading Magnetic due to the three coefficients discussed is:

A + B sin Hdg + C cos Hdg and the sign of the deviation will take care of itself.

e.g. A + 3, B + 2, C − 10

What deviation would be expected, (a) on 222(M), (b) on 146(M)?

(a)   + 3 + (+2 sin 222) + (−10 cos 222)
= +3 + (2 sin −42) − (10 cos −42)
= +3 − 2 sin 42 + 10 cos 42
= +3 − 1.3382 + 7.431
= 9° Easterly

(b)   +3 + (+2 sin 146) + (−10 cos 146)
= +3 + 2 sin 34 + (−10 cos −34)
= +3 + 1.11838 + 8.2904
= +12° or 12° East

Now to swing a compass: the methods are legion, and the usual method is to have a bod with a landing compass mounted accurately on a tripod well in front of the a/c so that he can sight down the fore and aft line: not as difficult as it sounds, since the tail unit can be sighted as a thin line in the viewer of the landing compass. This ineffably tedious job is now entirely in the hands of a trained Compass Adjuster, and the pilot only gets nobbled to drive occasionally. The procedure is as follows:

1. Check compass for serviceability (details of this under the title 'P type' compasses).
2. Ensure all equipment not carried in flight is removed.
3. Ensure all equipment carried in flight is correctly stowed.
4. Take the a/c to a suitable site, at least 50 yards from other aircraft, 100 yards from the hangar. Most airfields have a favourite and magnetically clean area for compass swinging.
5. Ensure that all flying controls are in normal flying position, engines on, radios and electrical circuits on.
6. Place a/c on Hdg South (M) and note deviation.
7. Place a/c on Hdg West (M) and note deviation.
8. Place a/c on Hdg North (M) and note deviation.
9. Calculate Coefficient C, apply it direct to the compass reading, and set the required corrected reading on the grid ring.
10. Place the Key across the needle and turn the Key until red is on red. (The Key is turned anti clockwise for + deviation.)

11. Remove the Key.
12. Place a/c on Hdg East, note deviation.
13. Calculate Coefficient B, and correct as before. (Key would be placed in fore and aft position.)
    The correcting swing is complete.
14. Carry out a check swing on 8 headings, starting with SE.
15. Calculate Coefficient A. Loosen retaining screw on rear bracket of compass, and turn the Instrument clockwise if A is + ve, by the quantity found.
    An example:

Correcting swing

	Landing Compass	P4	Deviation	Corrected Reading
S	184	183	+1	180
W	272	268	+4	270
N	000	353	+7	356
E	088	088	0	086

$$\text{Coefficient C } \frac{+7-1}{2} = +3$$

$$\text{Coefficient B } \frac{0-4}{2} = -2$$

and the corrected readings after these coefficients have been eliminated are as in the end column.

Check swing

L/C	P4	Devn	Residual Devn After Correction	
134	129	+5	+1	
181	177	+4	0	
225	222	+3	−1	
272	270	+2	−2	Coefficient A
314	308	+6	+2	$\frac{+31}{8} = +4$
357	353	+4	0	
048	044	+4	0	
094	091	+3	−1	

Now a deviation card is filled in to be placed next to the compass, so that who-ever takes the aeroplane can fly as accurately as possible the required Hdg(M).

FOR	STEER	
000	000	It is necessary in flight to check the
045	045	compass by astronomical means,
090	091	especially on freighters, when unusual
135	134	loads may be carried: and we've already
180	180	seen that soft iron magnetism does
225	226	change with change of magnetic latitude.
270	272	
315	313	

Occasions when a compass should be swung
1. New compass fitted.
2. Every three months.
3. After a major inspection.
4. With any change of magnetic material in the aircraft.
5. If transferred to another base involving a large change of latitude.
6. After a lightning strike or after flying in static.
7. After standing on one heading for more than 4 weeks.
8. At any time when the compass or recorded deviation is suspect.

Change of latitude and its effect on compass deviation
The conclusions arrived at so far are that the deviations leading to the resolution of
Coefficients B and C are horizontal hard iron P and Q, and vertical soft iron c and f.

Horizontal hard iron (HHI)
By definition, the deviating force is constant as measured, but with a change of
strength of the Earth's horizontal component H, the effect of this deviating force will
vary as the directive force on the compass needle varies. As H increases in strength,
so the effect of a constant HHI deviating component will decrease, and vice versa.
The deviation of the compass will be doubled if H is halved. This can be stated as

$$\text{Deviation due to HHI is proportional to } \frac{1}{H}$$

The value of H anywhere in the world is known, measured in Gauss, and a
directive force on the needle of .25 Gauss is resisting a deviating force more strongly
than one of .15 Gauss. The deviation can be resolved then at a new magnetic latitude
by the formula:

$$\frac{\text{New deviation}}{\text{Old deviation}} = \frac{\text{Old H}}{\text{New H}}$$

Vertical soft iron (VSI)
Once again, deviation must vary inversely as H, the ability of the needle itself to
resist any deviating force. Additionally, the magnetism induced into vertical soft iron
will vary as the vertical component of the Earth's magnetic field varies, i.e. it varies
as Dip. At the Poles, then, maximum deviation will be caused by VSI where maximum

Dip induces maximum magnetism: at the Magnetic Equator, where Z is nil, the value of induced magnetism in VSI will be nil.

Deviation due to VSI is proportional to Z and inversely proportional to H

$\therefore$ Deviation due to VSI is proportional to $\dfrac{Z}{H}$ or tan Dip

Dip, too, is known world wide, as an angular measurement from the horizontal; while Z is measured in Gauss, and Z changes its sign as it crosses the Magnetic Equator.

The new deviation due to VSI with change of magnetic latitude can be found from the formula.

$$\frac{\text{New Devn}}{\text{Old Devn}} = \frac{\text{New Tan Dip}}{\text{Old Tan Dip}}$$

Providing clarity is preserved between the two formula, a new deviation can be readily found when magnetic latitude has been changed.

At BERKER, where H = .15 gauss, deviation due to P is +3, and due to c − 4; Dip is 48°, what is total deviation due to P + c at POOTLE where H = .6, dip 32°?

HHI $\qquad \dfrac{\text{New Devn}}{\text{Old Devn}} = \dfrac{\text{Old H}}{\text{New H}}$

$\qquad\qquad \dfrac{x}{+3} = \dfrac{.15}{.6}$

$\qquad\qquad = +.75$

VSI $\qquad \dfrac{\text{New Devn}}{\text{Old Devn}} = \dfrac{\text{New Tan Dip}}{\text{Old Tan Dip}}$

$\qquad\qquad \dfrac{x}{-4} = \dfrac{\text{Tan } 32}{\text{Tan } 48}$

$\qquad\qquad = -4 \tan 32 \cot 48$

$\qquad\qquad = -2.27$

$\qquad\qquad$ Total devn due to P + c = −1.52 at POOTLE

Component R

This is vertical hard iron assumed for analysis to be situated above or below the compass. In level flight, its effect on the compass is nil, but tail down or up, the hypothetical magnet is resolved into two parts, one of which introduces a fore and aft effect, as it were a false component P.

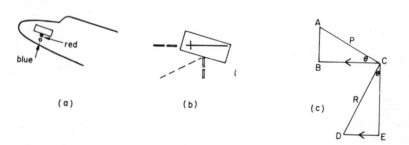

(a)   (b)   (c)

Fig. 20.6

In fig. 20.6(c) components P and R are shown (we can assume the inclusion of 'c') where AC = P and CD = R, $L\theta$ = the angle of tail down: the needle remains in the horizontal plane by construction.

Then BC = deviating effect due to P = P cos θ

DE = deviating effect due to R = R sin θ

∴ Total deviating effect = P cos θ + R sin θ; the action in the horizontal plane of component R introducing an additional effect like B (or P).

R is termed + with a blue pole beneath the compass and its deviative value depends on the angular distance of the a/c from the horizontal. It can be readily proved that the actual position of the applied force will change from fore to aft of the needle with change of a/c attitude from climb to descent.

e.g.   In the air, B is −7; on the ground 15° tail down, B is −4. What is the deviation on a Westerly heading climbing at 20°?

Total deviating force = B cos θ + R sin θ

$$-4 = -7 \cos 15 + R \sin 15$$
$$R = +11$$

In the air,

Deviation = −7 cos 20 + 11 sin 20

= −3, and this will be on East.

∴ +3° on Westerly Heading.

And another:

Climbing at 10° on East, the deviation due to P and R is +10°, whereas climbing on the same Heading at 20°, the deviation due to P and R is +3°. What is the deviation due to P on Hdg 315(M) in level flight?

Total deviation = P cos θ + R sin θ

$$10 = P \cos 10 + R \sin 10$$
and $3 = P \cos 20 + R \sin 20$
∴ $10 = .98P + .17R$
$$3 = .94P + .34R$$

which solves into

$$P = +17$$

In level flight, on a Hdg(M) of 315, with R inoperative, the deviation due to P will be .7 of −17, or to put it another way, +17 sin 315, an answer of −12°.

Finally, if the aircraft is banked, there will be a resolved component due to R appropriate to the rules of Coefficient C; but the manoeuvre is not sufficiently prolonged or directionally important enough to be worth considering, especially if account is taken of the larger effects of turning errors on the basic magnetic compass.

Sundry tips and reminders

First, are you a bit shaky on how to find cos 324? sin 179? tan 139? Whatever the angle, apply it to 360 or 180 to get below 90: then for the quadrant it falls in, the rule is:

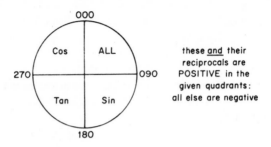

Fig. 20.7

For example:
　　　cos 324 = cos 036 and it's positive
　　　sin 179 = sin 001 and it's positive
　　　tan 139 = tan 041 and it's negative

Mathematics and so on

1.　Know the formula; $B = \dfrac{E - W}{2}$ and $C = \dfrac{N - S}{2}$

2.　If the value of B is known, the deviation on any other Heading is found from:
　　　Deviation = B sin Heading
　　Similarly, Deviation due to C = C cos Heading

3.　If the deviation <u>on a Heading</u> is given, B and C are calculated thus:
　　　$B = \dfrac{\text{Deviation}}{\text{sin Hdg}}$ 　　$C = \dfrac{\text{Deviation}}{\text{cos Hdg}}$
　　and these can be re-written:
　　　B = Deviation cosecant Heading
　　　C = Deviation secant Heading

4.　If the values of B and C are given and it is required to calculate the deviation on a given Heading, the problem is best solved in two parts: First, calculate the deviation due to B on the Heading given; then calculate the deviation due to C. The algebraic sum of the two is the deviation on that Heading due to the combined effect of both B and C.
　　　Deviation = B sin Hdg + C cos Hdg.

5.　<u>Maximum deviation</u>
　　Values of B and C are given, and it is required to find the Heading on which maximum deviation will occur, together with its value.
　　The formula to be used is:
　　　$\text{Tan Heading} = \dfrac{B}{C}$

This will give a Heading between 0 and 90 degrees. It will then be necessary to ascertain the quadrant in which the maximum value occurs and from knowledge of the quadrant, the Heading is converted into that quadrant. The quadrants are easily found from figs. 20.8(a), (b) and (c).

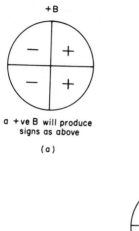

a +ve B will produce
signs as above

(a)

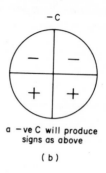

a -ve C will produce
signs as above

(b)

(c)

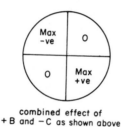

combined effect of
+ B and − C as shown above

Fig. 20.8

Example: Given B = +4, C = −3, find the Heading on which the deviation will be maximum.

From fig. 20.8, it is seen that the maximum Westerly deviation will occur between 270 and 360, and maximum Easterly deviation between 090 and 180.

$$\text{Tan Heading} = \frac{B}{C} = \frac{4}{3} = 1.333,$$ which is the Tan of 53° from the tables.

∴ Max westerly devn = 360 − 53 = 307

Max easterly devn = 307 − 180 = 127 (this must be the reciprocal of the Hdg which gave maximum deviation of opposite sign)

Once the Heading is known, the value of the deviation may be found from the normal formula, e.g. on Hdg 307

Total deviation = B sin Hdg + C cos Hdg
= 4(−sin 53) + (−3 cos 53)
= −4 × .7986 + (−3 × .6018)
= −3.1944 − 1.8054
= 4.999 Westerly

The value may also be found by Pythagoras, since B and C act at right angles to each other and we are interested in their resultant. In the above example:

(Maximum value)^2 = $4^2 + 3^2$
= 25

∴ max value = 5, and give the sign according to the quadrant.

The whole problem could be solved by scale drawing, and such a method is sometimes called for in the ALTP examination. But its real virtue lies in the fact that by learning this method, a firmer grasp of aircraft deviations is acquired. We'll do the same problem this way. It looks like fig. 20.9:

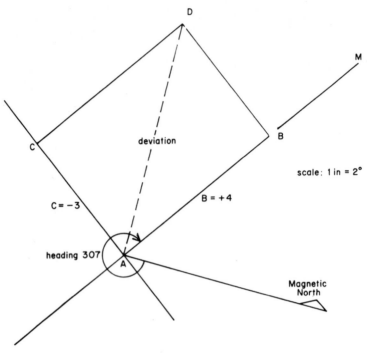

Fig. 20.9

Construction
 (i)   From a convenient point A, draw AM to represent any Heading of the aircraft.
 (ii) Along the Heading AM, let AB represent to scale the value of B, in this case +4. If it had been a −ve B, it would have been plotted towards the tail from A.
 (iii) At A and at right angles to AM draw AC to represent to scale the value of C, in this case −3. Being negative, C acts to the port of the aircraft.
 (iv) Complete the parallelogram ABDC. AD then represents the maximum deviation, and measures about 5°.
 (v)   To find the Heading on which this will occur, draw in the magnetic meridian from A, so that it is at right angles to AD, since maximum deviation occurs when the deviating force acts at 90° to the needle.
 (vii) Measure the Heading; it is 307.
 (vvi) The maximum Easterly deviation will occur on the reciprocal, 127.
5a. Maximum Deviation
     Another case is when the maximum value and the Heading on which it occurs are given, and it is required to separate B and C. You will recognise this as a problem in reverse of the previous one. It is best solved by scale diagram.

Example: Given that the maximum deviation is +5 on a Heading of 127°, find the value of Coefficients B and C.

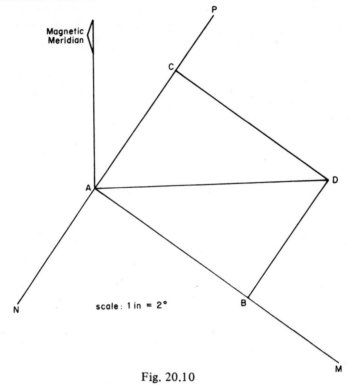

Fig. 20.10

In fig. 20.10, AM is the given Heading from the Magnetic datum, and NAP is the aircraft's athwartships axis.

AD is plotted to scale along the East-West line, and being given as +ve, in an Easterly direction therefore. The maximum value has in fact been plotted at 90° to the needle.

From D, drop two perpendiculars DB and DC to AM and AP respectively. Measure AB, 4°, and it is +ve since it acts towards the nose of the aircraft. Measure AC, 3°, and it is —ve since it acts towards the port. The same values for B and C would have been found if the given maximum deviation was —5 on a Hdg of 307; it's worth the effort of drawing out as a check.

6.   Zero Deviation

If the values of B and C are given, the Heading of zero deviation is found from the formula:

$$\text{Tan Heading} = \frac{C}{B}$$

and its derivation is as follows:

$$\text{Deviation} = B \sin Hdg + C \cos Hdg = 0$$

$$\therefore B \sin Hdg = C \cos Hdg$$

$$\frac{B}{C} = \frac{\cos Hdg}{\sin Hdg} \quad \text{or} \quad \frac{C}{B} = \text{Tan Hdg}$$

In the problem, tan Hdg = $\frac{3}{4}$ = .75 = 037° or 217°, the Headings where zero deviation occurs.

Of course, it will be appreciated that once the Heading of maximum deviation is found, the Heading of zero deviation is 90° removed from it. Occasionally, illustrations of zero deviation conditions are called for by means of a scale diagram.

<u>Example:</u> B is −ve and is half the value of C which is +ve. Draw a diagram to illustrate the heading on which these two forces will cause zero deviation.

The solution is similar to that for finding maximum deviation, except that the magnetic meridian is placed along the resultant and not at right angles to it.

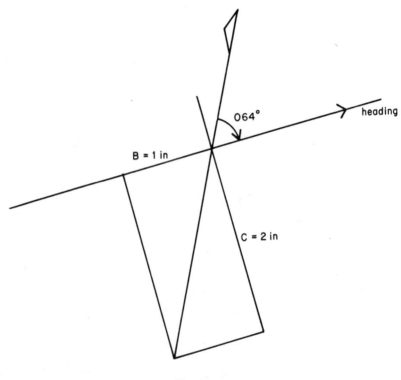

Fig. 20.11

Construction
(i)   Draw in any straight line to represent the aircraft's Heading.
(ii) At a convenient point on that straight line, draw in another straight line at right angles to it to represent the aircraft's athwartships axis.
(iii) From this point plot B and C. B is −ve, so is plotted toward the tail. C is +ve, so is plotted to the starboard.
(iv) Complete the parallelogram and extend the resultant as shown in the diagram to represent the Magnetic meridian.
(v)   Measure off the Headings of zero deviation. They are 064° and 244°.

21: Aircraft Magnetism 2

We learnt the theory of aircraft magnetism in the previous chapter. We will now consolidate it by working a few problems. At the start we summarise the properties of aircraft magnetism which should be kept firmly in mind when doing the problems.

	Hard iron (P and Q)	*Vertical soft iron (cZ and fZ)*
Strength	Hard iron is of permanent nature and therefore there is no change in its strength either due to the change of heading or the change of latitude.	The value of Z increases as the latitude increases; decreases as the latitude decreases; it is zero at the magnetic equator.
Polarity	There is no change in the polarity of hard iron either with change of heading or change of latitude.	The polarity reverses on crossing the magnetic equator.

Deviation

   Effect of change of heading

P varies as sine of the heading	cZ varies as sine of the heading
Q varies as cosine of the heading	fZ varies as cosine of the heading

   Effect of change of latitude

No effect on hard iron. Earth's component H varies with latitude,

$$\therefore \text{deviation} = \frac{1}{H} \text{ and}$$

$$\frac{\text{New dev}}{\text{Old dev}} = \frac{\text{old } H}{\text{new } H}$$

VSI is zero at the magnetic equator, maximum at the poles.

$$\frac{\text{New dev}}{\text{Old dev}} = \frac{\text{new} \dfrac{Z}{H}}{\text{old} \dfrac{Z}{H}} = \frac{\text{new tan dip}}{\text{old tan dip}}$$

Terminology. 'Landing compass reading' means 'heading magnetic'. Similarly, 'aircraft compass reading' means 'heading compass'. A plus (+) deviation is the same as easterly deviation and a minus (−) deviation is the same as westerly deviation.

Compass swing problems

1. An aircraft is swung on 4 cardinal headings with the following results. Calculate coefficients *A*, *B* and *C*.

Landing compass	Aircraft compass
000	001
089	092
176	175
271	270

Solution

The first step is to work out the deviation on all four headings. If you have any doubt as to the sign (+ or −) of deviations the golden rule is: make your aircraft compass read magnetic heading by adding or subtracting from the compass heading. That is your deviation. In the above problem make 001 read 000 by subtracting 1; the deviation is −1. Thus, we have:

Landing compass	Aircraft compass	Deviation
000	001	−1
089	092	−3
176	175	+1
271	270	+1

Coefficient A is found by adding algebraically the deviation on all 4 headings (or 8 headings if it is an eight-point swing), and by dividing this total by the number of points on which deviation was checked, (that is, 4 or 8).

Therefore,
$$A = \frac{-1 + (-3) + 1 + 1}{4} = -\frac{2}{4} = -\tfrac{1}{2}$$

Coefficients B and C are found by the formulae

$$B = \frac{\text{dev on east} - \text{dev on west}}{2} = \frac{-3 - (+1)}{2} = -\frac{4}{2} = -2$$

$$C = \frac{\text{dev on north} - \text{dev on south}}{2} = \frac{-1 - (+1)}{2} = -\frac{2}{2} = -1$$

Answer: $B = -2; C = -1; A = -\tfrac{1}{2}$.

2. Calculate coefficients B and C

Hdg(M)	Hdg(C)
359	357
091	089
180	180
271	273

Answer: $B = +2, C = +1$.

3. Calculate coefficients B and C

Hdg(C)	Hdg(M)
357	358
090	087
182	179
268	270

Answer: $B = -2.5;$ $C = +2$.

4. Calculate coefficients A, B and C

Landing compass	Aircraft compass
000	356
045	045
089	096
137	145
178	184
225	224
269	264
313	311

Answer: $A = -1°$; $B = -6°$; $C = +5°$.

5. Calculate coefficients A, B and C

a/c compass	Landing compass
001	003
048	043
095	087
136	133
181	175
224	225
268	274
313	317

Answer: $A = -1°$; $B = -7°$; $C = +4°$.

6. A compass is found to have following coefficients: $A = +1, B = -3, C = -2$. Determine the deviation on heading $210°(C)$.

Solution

The formula for calculating deviations when coefficients are known is:

$$\text{Dev on hdg } \theta = A + B \sin \theta + \cos \theta.$$

Thus,

$$\begin{aligned} \text{Dev on } 210° &= +1 + (-3 \sin 30°) + (-2 \cos 30°) \\ &= +1 + (-3 \times -0.5) + (-2 \times -0.86) \\ &= +1 + 1.5 + 1.7 \\ &= +4.2° \end{aligned}$$

Alternative layout:

$A =$	+1.0
$B = 3 \sin 30° = 3 \times 0.5 =$	1.5 (ignore the sign for time being)
$C = 2 \cos 30° = 2 \times 0.86 =$	1.7 (ignore the sign for the time being)

Total deviation (when all signs are known) ——?——

To determine the signs draw up two circles, one representing coefficient B and the other, coefficient C. Mark off roughly $210°$ points on both circles. Then look up the question again for the signs of B and C. The signs given are correct as follows:

Coefficient B: the sign is correct for easterly headings.

Coefficient C: the sign is correct for northerly headings.

Divide the circle in two halves as shown in the fig. 21.1. Coefficient B is -3, therefore, we put a minus (−) sign in the eastern half of the circle, a plus sign in the western half. Note that your heading mark falls in the plus (+) half. So, plus is the

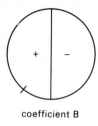

coefficient B

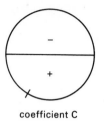

coefficient C

Fig. 21.1

sign for B. Insert the sign in the above table.

The coefficient C is negative, therefore a minus (−) sign goes in the northern half, a plus sign in the southern half. The heading mark is in the plus half, so the sign for C is, again, plus.

Thus, the total is: $+1° + 1.5° + 1.7° = 4.2°$.

7. $B = +7$, $C = −6$; what is the deviation on heading 200°(C)?
Answer: +3.25°.

8. $A = −1$; $B = −2$, $C = −3$; what is the deviation on heading 295°(C)?
Answer: −0.46°.

9. $A = −1$, $B = +2$, $C = −2$; what is the deviation on heading 315°(C)?
Answer: −3.8°.

10. $A = +2$, $B = −2$, $C = +3$; calculate the total deviation when the heading indicated by the aircraft compass is 060°(C)?
Answer: +1.8°.

Adjustment of compass

A swing may start on a northerly or southerly heading. If it starts on the north the aircraft heading is compared with the landing compass, the deviation is ascertained for north and then the aircraft is turned on to an easterly heading. The procedure is repeated and the aircraft is then taken on to south. Here we now have sufficient data to calculate coefficient C: that is,

$$\frac{\text{Dev N} - \text{Dev S}}{2}.$$

Let us say that we have these deviations:

a/c comp	L/C	Dev
004	001	−3
175	180	+5

These readings give the coefficient C of −4°. Since we are already on a southerly heading we might as well correct for C here. What must be remembered here is that the sign of the coefficient C is valid on north; therefore if the correction is made on south the sign must be reversed.

The aircraft is heading on 175°; coefficient is −4°, the correction to apply is +4; therefore, the compass must be made to read 179°. This adjustment is made by use of the compass corrector key in conjunction with the micro-adjuster mounted on the compass.

Coefficient C is corrected when the aircraft is heading north or south.

Positive (+) C is corrected by turning the key in one of the two keyholes in the athwartships axis in an **anti-clockwise direction.**

Negative (−) C is corrected by turning the key in one of the two keyholes in the athwartships axis in a *clockwise* direction.

Thus, in our case the key is turned clockwise until the compass reads 179°.

Had the swing commenced on south the correction for C would be made when heading north. The coefficient would then be applied without change of sign, that is, the compass is made to read: $004° - 4° = 000°$.

Now on to west and let us assume the deviation to be:

a/c comp	L/C	Dev
085	087	+2
274	270	−4

to give us the coefficient B of +3, which when corrected on a westerly heading, must have the sign reversed to obtain the correct required reading i.e. 271°.

To correct positive (+) B insert the key in one of the two keyholes in the fore-and-aft axis and turn in an anti-clockwise direction.

To correct negative (−) B insert the key in one of the two keyholes in the fore-and-aft axis and turn in a clockwise direction.

In our illustration the key is turned anti-clockwise until the compass reads $274° - 3° = 271°$. If the swing had commenced on south the correction would be made when on an easterly heading and the corrected compass heading would be $085° + 3° = 088°$.

Coefficient A may be corrected on any heading whatsoever and there is no question of any sign change since A is a constant value (a straight line curve) deviation throughout 360°. Thus, for example if A is −3°, then if adjusting on heading 212° you make the compass read 209°; on 050° you make it read 047° and so on. Coefficient A arises mainly due to the misaligned lubber line (although there is some A due to magnetic causes) and the correction is purely mechanical. Fig. 21.2 illustrates the cause of A due to a misaligned lubber line and explains the following rules of correction.

To remove positive (+) A, turn the compass bowl clockwise.

To remove negative (−) A, turn the compass bowl anti-clockwise. This can be seen in fig. 21.2 which illustrates a −ve A.

11. An aircraft compass was swung on 8 points with the following results.

a/c comp	L/C
000	003
046	044
090	084
135	129
179	174
227	223
270	270
315	319

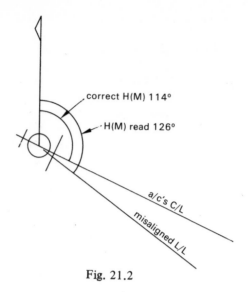

correct H(M) 114°

·H(M) read 126°

a/c's C/L

misaligned L/L

Fig. 21.2

(a) Calculate coefficients A, B and C.
(b) If B is compensated on east, what would you make the compass read?
(c) If C is corrected on south, what would you make your compass read?
(d) If A is corrected on 315°, what would you make your compass read?
(e) List the residual deviations that you would expect after compensation for coefficients A, B and C.

a/c comp	L/C	Dev
000	003	+3
046	044	−2
090	084	−6
135	129	−6
179	174	−5
227	223	−4
270	270	0
315	319	+4

$$A = \frac{+7 - 23}{8} = -\frac{16}{8} = -2°$$

$$B = \frac{-6 - (0)}{2} = -\frac{6}{2} = -3°$$

$$C = \frac{+3 - (-5)}{2} = +\frac{8}{2} = +4°$$

(a) $A = -2°$, $B = -3°$, $C = +4°$.

(b) a/c hdg = 090°
Coeff B = −3°
Make compass read = 087°

(c) a/c hdg = 179°
Coeff C = −4° (change sign)
Make compass read = 175°

(d) a/c hdg = 315°
Coeff A = −2°
Make compass read = 313°

(e) In a magnetic compass there are altogether five coefficients: *A, B, C, D* and *E*. Coefficients *D* and *E* are not in your syllabus but when you zero *A, B* and *C*, you still have some residual deviation left in the compass due to *D* and *E*. To work this question quickly and without getting bogged down in unnecessary calculations, the following layout is recommended.

(i) Make columns under the following headings and enter the information available from the question: aircraft headings and landing compass readings. The aim is to complete all the columns for all eight headings.

(ii) Columns under Hdg(C) and Hdg(M) will give you the deviations; enter them in the third column.

(iii) Now complete the column under *A*; it's the same number all the way through with the same sign: −2° in our illustration.

(iv) Column under *B*. On north and south the value of *B* is 0; on east, it is the value calculated (−3°); on the west, enter +3°. As for the deviations on the quadrantal headings, they are all functions of sine of 45°. The value to enter in our case is 3° x 0.7 = 2.1°. Having entered 2.1° on all quadrantal points, determine the sign from the deviation circle.

(v) Repeat the process for the figures under *C*. North is +4°, south is −4° and 0 on the east and west. The cosine of 45° is also 0.7 and the value to enter for quadrantal headings is 4° x 0.7 = 2.8°, the signs extracted from the circle for *C*.

(vi) Add the values of *A, B* and *C* for each heading and complete the next column 'total'. The figure in this column under each heading indicates the amount by which each heading was corrected when we compensated for *A, B* and *C*. Thus, on north: *A* = −2°, *B* = 0, *C* = 4°, Total +2°, and this is the correction applied to the compass heading which is 000°.

(vii) After the compensation, the compass on north reads 000° + 2° = 002°. This is entered in the column under 'compass reads'. Complete the column.

(viii) Finally the column under 'residual' is the answer. This is the remaining deviation after compensation. The values for this column are obtained by comparing 'what the compass reads now' with 'what it ought to read', that is, the magnetic heading. After the compensation on north the compass reads 002° whereas the landing compass reads 003°. The residual deviation for this heading is +1°. Complete the remainder of the column and your work should read as in the table below − the answer to the question (e) being the figures under 'residual'.

Hdg(C)	Hdg(M)	Dev	A	B	C	Total	Comp reads	Residual
000	003	+3	−2	0	+4	+2	002	+1
046	044	−2	−2	−2.1	+2.8	−1.3	044.7	−0.7
090	084	−6	−2	−3	0	−5	085	−1
135	129	−6	−2	−2.1	−2.8	−6.9	128.1	+0.9
179	174	−5	−2	0	−4	−6	173	+1
227	223	−4	−2	+2.1	−2.8	−2.7	224.3	−1.3
270	270	0	−2	+3	0	+1	271	−1
315	319	+4	−2	+2.1	+2.8	+2.9	317.9	+1.1

12. There follows the result of a compass swing.

Hdg(C)	Hdg(M)
001	004
045	047
091	088
134	130
181	176
226	224
271	270
315	317

(a) Determine the coefficients, A, B and C.

(b) Determine the residual deviations existing on all eight headings.

Answer: (a) $A = -1$; $B = -1$; $C = +4$

(b) $0, +0.9, -1, +0.5, 0, +1.1, -1, -0.5$.

13. There follows the result of an aircraft's compass swing.

Hdg(C)	Hdg(M)
002	357
047	044
092	090
137	135
182	181
227	228
272	272
317	313

(a) Give the values of the coefficients A, B and C.

(b) Explain how you would compensate for coefficients B and C.

(c) Explain how you would compensate for coefficient A.

Answer: (a) $A = -2$, $B = -1$, $C = -2$.

(b) Since the swing commenced on a northerly heading, the coefficient C will be the first to be corrected on a southerly heading.

a/c hdg	182°(C)
Coeff C	+ 2° (change sign)
Make compass read	184°(C)

This adjustment is made by inserting the micro-adjuster key in either of the key-holes in the athwartships axis and turning the key clockwise until the new heading is obtained.

Coefficient B is corrected on heading 272°(C)

a/c hdg	272°(C)
Coeff B	+ 1° (change sign)
Make compass read	273°(C)

To make this adjustment the compass key is inserted in either hole in the fore-and-aft axis and turned clockwise until the required heading is indicated.

(c) Coefficient A is usually corrected in the hangar after the field operation is complete. Therefore the aircraft may be on any heading when the correction is being made. Assuming the aircraft is on 220°(C). Coefficient A is −2°. Loosen the retaining

screws at the base of the compass and rotate the base anti-clockwise through 2°
against the scale at the base provided specifically for this purpose. The effect of this
is to rotate the lubber line anti-clockwise through the same amount, thus reducing
the heading. Set up the new reading against the lubber line (thus upsetting the red/red
relationship). After the bowl has been rotated through the requisite angle note that
red is again aligned with red. The new heading is then being indicated and coefficient
A has been removed. Fasten down the base screws.

14. During a swing of a direct reading magnetic compass the following readings were
obtained.

Landing compass	a/c compass
268	270
357	352
089	085
176	175

(a) Determine the values of coefficients A, B and C.
   (b) If coefficient B is corrected on heading 089°(C) which keyhole would you
use, and which way would you turn the key?
Answer: (a) A = +2, B = +3, C = +2.
   (b) Using either keyhole in the fore-and-aft axis, turn the key anti-clockwise.

15. An aircraft compass swing gives the following readings;

H(C)	H(M)
002	001
091	096
183	186

At this stage when the aircraft is on the southerly heading, coefficient C was
compensated.
   (a) What should the compass be made to read?
Assuming the coefficients present in the compass are A, B and C only
   (b) what deviation would you expect to find when on the compass heading
of 272°(C)?
   (a) 185°(C).
   (b) The only coefficients present are A, B and C. On a southerly heading, there
can be no deviation due to coefficient B; therefore, the only coefficients present are
A and C. After adjusting for C on south, we note that the aircraft heading is 185°(C)
whereas the magnetic heading is 186°(M). This difference is due to coefficient A, +1°.
   Before we arrive at the deviation on 272° we must establish the value of coefficient B. On east and west there can be no coefficient C. Therefore the only
coefficients present on these headings are A and B. On east, the deviation measured
is +5°. We know that we have +1° of coefficient A, therefore, deviation +5° is made
up as follows:

+1° of coefficient A
+4° of HI and/or VSI (P and/or cZ)

Therefore, +4° is the coefficient B we are looking for. Any other figure for coefficient
B may give you +5° deviation on the east, but not without producing a residual. This

may be checked in the following.

Hdg(C)	Hdg(M)	Dev	A	B	Total	Comp reads	Residual
091	096	+5	+1	?	——	096	0

In the above, we know that after compensating for A and B the residual must be 0 (there is no C on east and there are no other coefficients present). If the residual is to be 0, the compass, after the compensation, must read the same as the magnetic heading, which is 096°. This can only be achieved if the total compensation is +5° (091° + 5° = 096°). This +5° is the total of coefficients A and B. Therefore, if A is +1, B has to be +4°.

$$\text{Now, } B = \frac{E - W}{2} \quad \text{or} \quad 2B = E - W$$

and
$$W = E - 2B$$
$$= +5° - 8°$$
$$= -3°$$

So the answer is −3°, but you need not go through this process of shifting the equation. Notice that B is +4°; on the west it changes the sign: −4°; to this apply the coefficient A to give you the total deviation, which is −3°.

Change of deviation with latitude
Fig. 21.3 shows the effect on deviations due to a change of latitude, where the deviations are caused by hard iron only. A hard iron component, *HI*, is shown acting

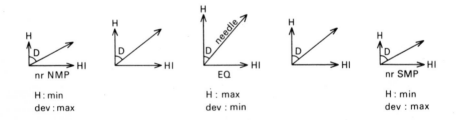

Fig. 21.3

at right angles to the directive force, H: firstly, in the very close vicinity of the two poles, secondly, at intermediate latitudes in both the hemispheres and thirdly at the magnetic equator. Note from this diagram that

(a) the magnitude of the hard iron present in the aircraft does not change with latitude. The *HI* vectors shown in the diagram are of a constant length.

(b) The earth's component H is least at the vicinity of the poles, and greatest at the equator. The needle takes up the position resultant of these two vectors, causing changes in deviations. And since *HI* remains constant throughout, the amount of deviation at any place is dependent upon the strength of the component H only.

The relationship is

$$\tan D = \frac{HI}{H} \quad \text{and since } HI \text{ is constant}$$

$$D \propto \frac{1}{H}$$

This is the basic relationship. When we connect the deviations at two latitudes (and that's what we are interested in) the formula becomes:

$$\frac{D_1}{D_2} = \frac{H_2}{H_1}$$

and from this since we always seem to want the deviation at the second place, D_2,

$$D_2 = \frac{D_1 H_1}{H_2} \qquad \text{(hard iron formula)}$$

(c) In fig. 21.3 the deviation is easterly all the way through. Although with change of latitude the value of deviation changes, the sign of the deviation does not change. If the deviation due to HI at X was negative, it would still be a minus quantity at Y no matter where Y was situated.

Now for the vertical soft iron, VSI. VSI is induced by the component Z of the earth's field. Z varies with magnetic latitude, growing weaker away from the magnetic poles, and the induced fields similarly vary in proportion. Therefore, if an aircraft's compass deviations are caused entirely by VSI, both the vectors that influence the needle, that is, H and VSI are variable with change of latitude. The effect of these two variables on deviation is shown in fig. 21.4. Nearer the two magnetic poles the

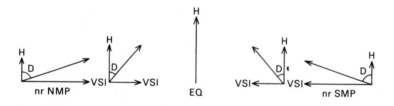

Fig. 21.4

value of Z is very large, the value of H very small with consequent very large deviations. Approaching the magnetic equator, Z decreases, H increases until the magnetic equator is reached where Z completely disappears, H is maximum and no deviations are present. Note from fig. 21.4 that on crossing the magnetic equator from one hemisphere to another the VSI reverses its polarity. (In fig. 21.5 we show why this occurs.) From fig. 21.4,

$$\tan D = \frac{cZ}{H} \quad \text{and} \quad D \propto \frac{Z}{H}.$$

Since $\frac{Z}{H}$ also equals tangent of dip, the formula linking the deviations at two places may be written down in two ways:

(a) $\quad \dfrac{D_1}{D_2} = \dfrac{\text{Dip}_1}{\text{Dip}_2} \qquad$ (vertical soft iron formula)

(b) $\dfrac{D_1}{D_2} = \dfrac{Z_1/H_1}{Z_2/H_2}$; reversing this: $\dfrac{D_2}{D_1} = \dfrac{Z_2/H_2}{Z_1/H_1}$ and

$$D_2 = D_1 \times \frac{H_1}{Z_1} \times \frac{Z_2}{H_2}$$

$$D_2 = \frac{D_1 H_1 Z_2}{H_2 Z_1} \qquad (VSI \text{ formula}).$$

You have perhaps noted that the hard iron formula is multiplied by Z_2/Z_1 to arrive at the *VSI* formula.

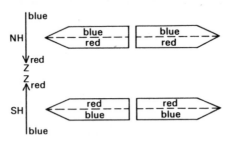

Fig. 21.5

In fig. 21.5 vertical soft iron is shown on easterly and westerly headings in both hemispheres. The earth's component Z is shown acting downward in the northern hemisphere and upward in the southern hemisphere. This component Z induces magnetism in the *VSI*. Note from fig. 21.5 that

(a) in the northern hemisphere the lower end of the *VSI* becomes red, upper end blue.

(b) In the southern hemisphere the lower end is blue, the upper end red. Thus there is a reversal of polarity with change of hemisphere.

(c) Note also that there is no change of polarity with change of heading (east and west polarities). This fact enables us to combine *VSI* with *HI* in calculating coefficients B and C.

In dealing with the problems where the compass was left completely uncorrected, you simply calculate the new deviation using the *HI* and *VSI* formulae as appropriate.

If the compass is corrected at X, calculate the deviation at Y as follows:

(a) If it was entirely due to *HI*, and completely compensated, then the deviation at X is 0 and the deviation at Y will also be 0. If it was partially compensated, say $2°$ out of $+4°$, then you have $2°$ of easterly deviation uncompensated and treat this as if the compass has not been corrected.

(b) Total or partial correction of *VSI* holds good only for the particular values of H and Z. Therefore, where the deviation is entirely due to the *VSI* and corrected, the new deviation can be calculated in two steps:

  (i) Calculate the effect of the micro-adjuster at Y using the hard iron formula and remembering that this *HI* of the micro-adjuster is of opposite polarity to the deviation it corrected at X. Thus, if the deviation at X was $+2°$, then when correcting for this, you inserted a micro-adjuster *HI* polarity of $-2°$ to neutralise it.

(ii) Then calculate the dev due to *VSI* at Y, using the *VSI* formula.
The sum of the two above is the new deviation.

(c) If the deviation at X is due to both *HI* and *VSI*, do the problem in three steps:

(i) Calculate the micro-adjuster effect on Y using the *HI* formula;

(ii) Calculate the new deviation due to *HI* part using the *HI* formula;

(iii) Calculate the new deviation due to *VSI* part using the *VSI* formula.

The sum of the three is the new deviation at Y.

Latitude change problems

1. The values of H and Z at two places X and Y are these:

	H	Z
X	0.4	0.3
Y	0.3	-0.2

The coefficient B at place X is made up of *HI* of $+2°$ and *VSI* of $-2°$. What is the value of coefficient B at Y?

Solution

$$\text{Dev due to } HI \text{ at } Y = \frac{D_1 H_1}{H_2} = \frac{+2 \times 0.4}{0.3} = +\frac{8}{3} = +2.66°$$

$$\text{Dev due to } VSI \text{ at } Y = \frac{D_1 H_1 Z_2}{H_2 Z_1} = \frac{-2 \times 0.4 \times -0.2}{0.3 \times 0.3} = \frac{16}{9} = +1.77°$$

$$\text{Total dev at Y (coeff } B) = +4.43°$$

2. At position X ($H = 0.3$, $Z = 0.3$) the coefficient B of $+1\frac{1}{2}°$ is entirely due to vertical soft iron. If this has been entirely compensated at X, determine the deviation on heading $315°$(C) due to coefficient B at Y ($H = 0.15, Z = -0.4$)
Answer: $+4.9°$

Solution
Coefficient B of $+1.5°$ has been compensated by means of a micro-adjuster. Therefore, at X, a hard iron of $-1.5°$ has been introduced, and although at X, it neutralised the *VSI*, at Y it will produce deviation of its own.

$$\text{Micro-adjuster: } D_2 = \frac{D_1 H_1}{H_2} = \frac{-1.5 \times 30}{15} = -3°$$

$$VSI: D_2 = \frac{D_1 H_1 Z_2}{H_2 Z_1} = \frac{+1.5 \times 30 \times -4}{15 \times 3} = -\frac{12}{3} = -4°$$

$$\text{Coefficient } B = -7°$$

$$\text{Dev on } 315° = B \sin 315°$$
$$= -7° \times (-\sin 45°)$$
$$= +4.9°$$

Max min deviations, polarity, relative bearings

When dealing with these problems keep these points firmly in your mind.

(a) If a blue or red pole is located in the fore-and-aft axis, it can only produce coefficient B. Similarly, poles lying in the athwartships axis produce only coefficient C. A pole lying in the intermediate position will produce both B and C coefficients.

(b) When the aircraft (or deviascope) turns, these poles which are fixed in the frame turn with the aircraft, and therefore their relative position with regard to the needle alters, altering the deviation.

(c) When the blue pole is aligned with the needle, maximum directive force results. When the red pole is aligned with it, minimum directive force results.

(d) When the poles are located at 90° to the needle, maximum deviation results.

1.   A vertical soft iron bar is placed on a deviascope, fore of the fore-and-aft axis and produces a coefficient of magnitude 4. The compass is in the southern hemisphere and the lower end of the bar is level with the magnet system. What deviation will be caused by this bar when on heading 300°(C)?

Solution
The problem basically involves identification of the pole giving a coefficient of 4.

Since the bar is placed in the fore-and-aft axis, the coefficient is B. And since the lower end of the bar is level with the compass and the compass is in the southern hemisphere, the polarity in front of the needle is blue.

And lastly, a blue pole in line with the needle in the fore position in the fore-and-aft axis gives a positive coefficient B.

Now the problem is solved in the normal way.

$$\text{Dev}_{300} = B \sin 60°$$
$$= +4° \times (-0.87)$$
$$= -3.48°$$

2.   A hard iron source on a deviascope produces a maximum easterly deviation of 6° when on a heading of 315°(C).

Show the locations of the blue and red poles causing this deviation in a diagram.

Solution

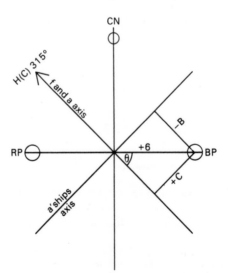

Fig. 21.6

In fig. 21.6 the poles have been placed at 90° to the needle as the deviation is maximum. Since the deviation is easterly, in order that the needle be pulled clockwise, the blue pole is placed to the right of it, red pole to the left.

3.   For the data in question 2, what is the deviation on heading 210°(C)?

Solution
To be able to calculate the deviation on 210°(C) (or any other heading for that matter) we must have both the value and the signs of coefficients B and C.
   The values of B and C are worked out by trigonometry. In fig. 21.6, drop perpendiculars from the blue pole to the fore-and-aft and athwartships axes of the aircraft. These represent coefficients C and B respectively.
   Since we know the aircraft heading we are able to work out the value of angle θ in the figure. This angle is 45°. Now we have sufficient information to calculate B and C.

$$\text{Sin } 45° = \frac{C}{+6°} \quad \text{and} \quad C = 4.2°$$

$$\text{Cos } 45° = \frac{B}{+6°} \quad \text{and} \quad B = 4.2°$$

As for the signs, looking at the figure once again, we notice that in the fore-and-aft axis B is acting aft of the compass position in the fore-and-aft axis, and therefore it is −ve B. Similarly, C is acting to the right of the centre line in the athwartships axis; therefore, C is positive. Thus, the coefficients are

$$B = -4.2°$$
$$C = +4.2°$$
$$\text{Dev}_{210} = B \sin 210° + C \cos 210°$$
$$= (-4.2° \times -0.5) + (+4.2° \times -0.87)$$
$$= +2.1° - 3.7°$$
$$= -1.6°$$

4.   For the data in question 2, on what heading is the directive force (a) maximum? (b) minimum?

Solution
   (a) For maximum directive force, the blue pole must lie in front of the red point of the needle. Referring to fig. 21.6, the blue pole will come in this position if the heading is altered to port through 90°. Thus, the heading for maximum directive force = 315° − 90° = 225°(C).
   (b) For minimum directive force, the red pole must lie against the red point of the needle. This heading is 225° − 180° = 045°(C).

5.   Given: $B = +3°$, $C = +2°$.
   (a) On what heading is the directive force maximum?
   (b) On what heading is the directive force minimum?
   (c) On what heading is the deviation maximum?
Answer: (a) 326°19′; (b) 146°19′; (c) 236°19′ and 056°19′.

22: P-type Compass; E2 Compass

The basic compass is simply a needle which points to the North, adapted for aerial use. It is mandatory to carry in civil transport aircraft a compass which does just this, so that if the splendidly refined direction-indicating gear goes for a burton, you are down to basics. The 'P' type is rather bulky and is seldom selected nowadays by sophisticated operators, but it is in constant use world-wide nevertheless, and is beloved of examiners.

A freely suspended magnet would settle along the Earth's line of total force, providing it were clear of other magnetic influences, which means it would point to Magnetic North in the horizontal plane, and dip from the horizontal in the vertical plane (66° in the U.K.): the latter effect must be got rid of somehow, for we are interested only in the Earth's horizontal force H.

For the compass to function efficiently it must:

1. Lie horizontal
2. Be sensitive
3. Be aperiodic or dead-beat.

The first requirement is obtained by making the magnet system pendulous. Four magnets are used in fact, and they are mounted close together below the pivot, so that when tilted by the Earth's vertical force Z, the centre of gravity moves out from below the point of suspension bringing a righting force into action: the magnet will take up a position which is the resultant of the two equal and opposite forces, the tilting of the needle due to Z being counteracted by the weight of the magnet system acting through the C of G. The final inclination of the compass to the horizontal is actually about 2°/3°, but it will increase with increased Z until 70N or S where H is so weak and Z so strong as to render the instrument useless.

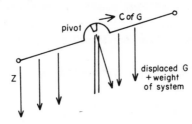

Fig. 22.1

In the N. hemisphere, the C of G would be considered to be southwards of the point of suspension, since the correcting force must be away from the north seeking end which is trying to dip: it must be stressed right now that this displacement of the C of G is a factor brought into play by the system's pendulosity: there is not, repeat not, some built-in mechanical adjustment to the pivoting of the needle which would render such a compass useless in the other Hemisphere. The C of G is displaced automatically from beneath the pivot to correct for dip whichever end of the needle does the dipping.

The second requirement of sensitivity can only be done by increasing the pole strengths of the magnets used so that the needle stays firmly fixed along the magnetic meridian. This is helped by keeping pivot friction to a minimum by using iridium for the pivot which is suspended in what is laughingly called a sapphire cup: it's made of corundum actually. All this is suspended in liquid which reduces the effective weight of the system and lubricates the pivot.

The third requirement of aperiodicity is a trifle more complex. If a suspended magnet is deflected from its position of equilibrium and released, it oscillates between positions on either side of the equilibrium position for some time before coming to rest. Period of oscillation is the time taken in seconds to travel from one extreme position to the other and back again. A period is undesirable in a compass, and the ideal compass would stop without oscillation, when it could be said to be 'aperiodic'. In the air, the compass needle is readily moved from its alignment with the magnetic meridian by accelerations and turns. Aperiodicity is attained in a/c compasses by the following means:

1. The bowl is filled with methyl alcohol, and damping filaments are fitted to the magnet system.
2. Several short powerful magnets are used instead of one large one, thereby increasing the righting force, and reducing the moment of inertia.
3. The apparent weight of the system is reduced by the buoyancy of the liquid, and that weight is concentrated near the centre, by mounting close to the pivot. The moment of inertia is even further reduced.

The liquid is actually methyl alcohol, double deaerated; ideally it should be transparent — it is; have wide temperature range — it has, −50° to +50°C; low coefficient of expansion — it hasn't, it's about 12% over the full temperature range; low viscosity and low specific gravity — not too bad; non-corrosive — not so important with plastic bowls, but it can eventually get at the rubber gasket of the verge ring. Since the liquid will change in volume with temperature variations, some expansion device is necessary. At the bowl base, an expansion chamber is fitted in the larger types or a sylphon tube in the smaller, of thin corrugated metal.

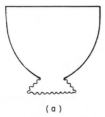
(a)

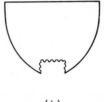
(b)

Fig. 22.2

One disadvantage of the use of liquid is that in a prolonged turn it will turn with the aircraft, taking the magnet system with it tending to affect indicated readings on completion. This is offset to some extent by keeping a good clearance between the damping wires and the wall of the bowl; the effect is small, since the viscosity of the liquid used is low, too, but liquid swirl does prevent an immediate settling down on a new heading compass.

The Magnet System

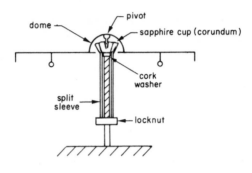

Fig. 22.3

The split sleeve overlaps the ledge surrounding the dome and prevents the whole works falling apart when the compass is inverted.

The bowl

The lubber line on the bowl is fixed on the a/c's fore and aft axis precisely, and we have already seen the effect of any displacement of it, as well as the ease of correction. The suspension of the bowl and its expansion device is such that vibration which would harm the pivot is reduced to a minimum: it stands on 4 bronze helical springs which are attached to the outer case spaced at 90° apart, and a floating ring, to which friction springs are attached, supports the bowl in slots to prevent azimuth movement.

The grid ring fits over the verge ring which seals the liquid in the bowl, and has direction marked every 10° graduated in 2° divisions, luminous. Additionally, the North point is a red triangle, hence the expression 'red on red', and parallel lines are marked across the plate, to be lined up with the needle precisely: the locking device is put to 'lock' once the desired heading is set against the lubber line, and the a/c turned till 'red is on red'. The North pointer of the needle is crossed to prevent error.

Serviceability checks

1. Check liquid is free from bubbles, discoloration, sediment.
2. Examine all parts for luminosity.
3. Ensure that grid ring rotates freely through 360°, and that locking device functions positively.
4. Test suspension of bowl by moving gently in all directions and that there is no metal to metal feeling.
5. Test for pivot friction: deflect the magnet system through 10–15° each way, and note the reading on return: each should be within 2° of the other.

6. Test for damping: deflect system through 90°, hold for 30 seconds to allow
    liquid to settle and time its return through 85°. The maximum and minimum
    times are laid down in the manufacturer's Instrument Manual, usually about
    6.5 seconds to 8.5 seconds.

The E2 compass

A small compass, vertical reading, which is the favourite as a standby compass:
invariably fitted between the two pilots above the windscreen, it is not expected to
be precise, but as a rough check on the main compass. It has correctors built in for
Coefficients B and C and R (there is another model, the E2A, which omits the R
corrector), on top of the bowl.

The bowl is of transparent plastic, and the lubber line is simply a luminous line in
the window: the magnet is a steel circle, domed, and from the dome a pivot is
dropped which rests in a sapphire cup on a stem fixed to the base of the bowl. Thus
dip is minimised as in the P type. The bowl is filled with a silicone fluid, non cor-
rosive, of low viscosity, and with a low coefficient of expansion — but there is a syphon
tube at the rear of the compass bowl.

The compass card is a light metal ring, marked off every 30°, graduated in 10°,
attached to the circular magnet, and the dome of the bowl prevents the system falling
off when inverted. The E2 suffers from turning and acceleration errors, and could in
extreme climb or dive positions be not free to rotate.

23: Turning and Acceleration Errors

We're still on about compasses where a magnet system is fitted pendulously to counteract the effect of Dip: the residual angle from the horizontal is around 3° in U.K. latitudes, and in the reduction from 66° to 3° the C of G has been displaced from below the pivot to South of the pivot to pull the dipping North-seeking end up towards the horizontal. This displacement will vary with Z, and clearly in the S. Hemisphere the displacement will be of opposite sign, i.e. the C of G will displace itself to the North of the pivot to counterbalance the dipping South seeking compass pole.

This displacement, however, has a marked effect on the compass needle in turns, accelerations and decelerations, except at the Magnetic equator where Z is nil, and a fine old song and dance is made about these errors to pilots on their way up.

Acceleration errors

On a Westerly heading, if speed is increased, the pivot and the magnet system will move forward with the a/c: there will also be an equal and opposite force acting on the centre of gravity, which in the horizontal plane is South of the Pivot: the resultant of this horizontal couple will rotate the needle in an anticlockwise direction, i.e. the North seeking end will be moved to the west. Additionally since the C of G is placed below the pivot in the vertical plane, the pivot assembly going forward while the C of G lags behind will cause a vertical tilt to the needle, since the needle, the pivot and the C of G are no longer in line with the magnetic meridian: the counteracting C of G against dip will thus be lost to some extent and the North end of the needle will be under some influence from Z, causing further rotation of the North seeking end in the direction of the acceleration. In fact, it is this part of the deal which contributes the major amount of the total error.

Fig. 23.1

Accelerating on an Easterly Hdg, with similar resultant forces, the needle would turn clockwise, showing an apparent turn to port, i.e, to the North. In all changes of speed, where the a/c's heading is across the needle there will be a compass error until constant speed is regained.

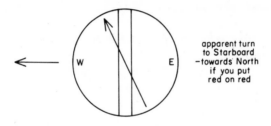

Fig. 23.2

Summary of Acceleration and Deceleration Errors

HDG	SPEED	NEEDLE TURNS	EFFECT
E	Increase	Clockwise	Apparent Turn to N.
W	Increase	Anticlockwise	Apparent Turn to N.
E	Decrease	Anticlockwise	Apparent Turn to S.
W	Decrease	Clockwise	Apparent Turn to S.

N.B.
1. In Southern Hemisphere, above errors are opposite
2. No errors on Nly and Sly headings as the force acts along the needle.
3. They can occur in bumpy conditions.
4. No errors on the Magnetic Equator, as the pivot and the C of G are coincident.

Turning errors
These are maximum on N and S headings, and are important within 35° of these headings.
   Consider a turn through N to starboard: the centripetal force acting on the pivot directed inwards towards the centre of the turn should balance the centrifugal force acting outwards: but the centrifugal force acts on the centre of gravity, which we know in the N. Hemisphere to be displaced south of the pivot. Thus the needle will be pulled outwards from its centre of gravity, turning the needle in the direction of turn, Easterly: and the compass will indicate less than the turn actually accomplished. Additionally, the unbalanced centrifugal/centripetal forces will set up a vertical pull on the slightly tilted North-seeking pole, accentuated by the fact that the needle has left the magnetic meridian; this vertical pull in the turn will manifest itself as a rotation of the magnet system in the direction of the turn, to the East: again, this is the major contribution to the error.
   In the turn we are considering, the needle can turn at the same rate as the a/c, and indicate no turn at all: it can turn slower than the a/c, indicating a smaller turn than actually accomplished: it can, in high rates of turn, move quicker than the a/c and indicate a turn in the wrong direction. In turns through South (we're in the N. Hemisphere) the reverse will take place, and the compass will rotate in the opposite direction to the a/c, indicating a greater turn than that actually made: but as the

needle turns as it were to meet the a/c, the result always is an overreading, and always in the right direction.

No fixed figure can be laid down to any of these errors, as so many variables are involved: but errors of up to 60° are not unusual in a prolonged turn, and only experience of a particular a/c with its particular compass type can avoid the fiddling and twiddling to get settled down on the precise compass heading required as soon as possible.

In the Southern Hemisphere, with the C of G displaced to the North of the pivot, the reverse results will be obtained. The swirl of the liquid which tends to turn in the same direction as the aircraft will, in the Northern Hemisphere, be additive to the error in turns through North, and subtractive in turns through South.

To summarise:

TURN	NEEDLE	EFFECT	LIQUID SWIRL	CORRECTION
Through North	Same as a/c	Under Indication	Adds to Error	Turn less than needle shows
Through South	Opp. to a/c	Over Indication	Reduces Error	Turn more than needle shows

N.B.
1. In Southern Hemisphere, above errors are of opposite value.
2. In turns about East and West, no errors to speak of, since forces act along the needle.
3. Northerly turning error is greater than Southerly, since liquid swirl is additive.
4. For accurate turns, use the DGI. How about that? All dynamic errors last only for the period of speed changes or turns: once a constant speed or level flight is resumed the compass needle finds North again.

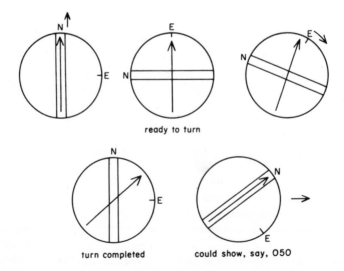

ready to turn

turn completed   could show, say, 050

Fig. 23.3

Check these pictures, for a turn to starboard from North, where the needle turns slower than the a/c. Repeat them for a variety of turns through N and S, in both hemispheres.

And for a rule of thumb:

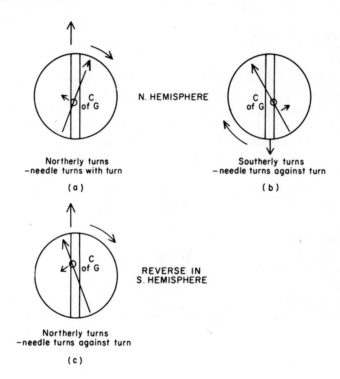

Northerly turns
−needle turns with turn

(a)

N. HEMISPHERE

Southerly turns
− needle turns against turn

(b)

REVERSE IN
S. HEMISPHERE

Northerly turns
−needle turns against turn

(c)

Fig. 23.4

24: Sperry's CL2 Compass

The CL2 is a remote indicating compass and combines the use of a gyro with the Earth's magnetic lines of force, thus giving the best of both DGI and P type compasses. As for the magnetic part the system does not align itself with the meridian; it senses it, as we shall soon see. Since it is a remote indicating compass, the detector unit which senses the meridian can be placed outside the cockpit. In fact it is usually tucked away in the wing tip or other remote part of the fuselage where the magnetic interference is the least. This gives it a high accuracy of $\pm\frac{1}{2}°$. The compass incorporates a facility to set variation, thus giving true headings, if required. Finally, the heading information may be transferred to a number of repeaters in the cockpit.

Principle of direction sensing

If you expose a highly permeable magnetic bar to the Earth's field the bar will acquire magnetic flux. The amount of flux thus produced will depend on two factors: latitude which governs the strength of the Earth's component H (assuming the bar is horizontal) and the direction of the bar relative to the direction of the component H. This second part is the key to the heading sensing.

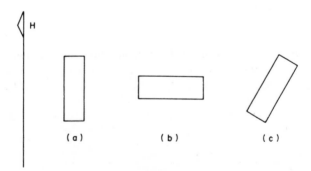

Fig. 24.1

In fig. 24.1(a) above a bar is placed having its axis directly in line with the magnetic meridian as defined by the direction H. Since H goes right through the length of the bar the flux produced is maximum. If this bar was fixed in the fore-and-aft axis of the aircraft, maximum flux will result when the aircraft is on heading 000(M). In fig. 24.1(b) the aircraft carrying the bar is on heading 090(M). Since component H acts at 90° to the axis of the bar, no flux will be induced. This is a cosine relationship. If the aircraft was on heading 030(M), the flux intensity would be H cos 30°.

Now, on heading 150(M) the flux intensity will be the same as on 030(M) but the direction of the flux flow will have reversed (cosine of 150° is negative).

This, therefore, gives us a basic principle which may be adapted to give direction measurements. The flux intensity is the measure of the heading. The problem is to convert flux into electrical voltage and current. Now, this is an easy matter if the flux so produced was 'changing' flux, for, according to Faraday, 'Whenever there is change of flux linked with a circuit an EMF is induced in the circuit'. It will be appreciated that in an aircraft at any given position the flux produced will be a steady, constant quantity. If this steady flux could be converted to changing flux, a current representing heading would flow. This is what we want, a current representing aircraft's heading.

Flux Valve. This is achieved in the CL2 by a device called the Flux Valve. It consists of two bars of highly permeable magnetic material – bars A and B in fig. 24.2.

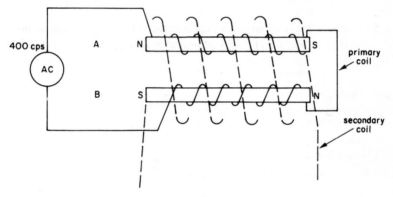

Fig. 24.2

Both the bars are wound with a coil and connected to an AC source in series. The coil is called the Primary Coil. Over the primary coil and going round both the bars is a pick-up coil, called the Secondary Coil. The effect of passing AC through these

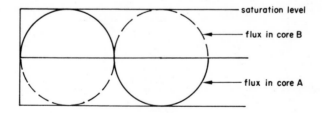

Fig. 24.3

wirings is shown in fig. 24.3. The AC passed is of such strength that at the peak it saturates both the coils, as seen in the figure. However, the flux produced in fig. 24.3 will have no effect on the secondary coil since at any given instant two bars produce flux of equal intensity and opposite polarity. But in practice this situation does not arise, since a bar placed horizontally on the Earth's surface or at an altitude (as in an aircraft) has always present in it the Earth's component H (unless the aircraft is flying in the close vicinity of either Pole). This component H produces a static flux in both the bars which is shown in fig. 24.4.

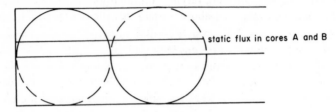

static flux in cores A and B

Fig. 24.4

The effect of this static flux is to saturate the cores before AC reaches the peak, as seen in fig. 24.5. The coils are saturated and AC is still rising. The effect will be

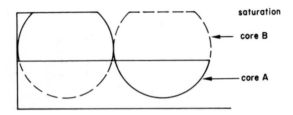

saturation

core B

core A

Fig. 24.5

that from now on, as the saturation occurs, the intake of the Earth's flux will start reducing and on a combined figure it will appear as a curve — a changing flux. Fig. 24.6.

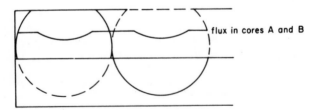

flux in cores A and B

Fig. 24.6

This changing flux will now produce voltage and current will flow in the secondary coil.

Basic components of the compass
CL2 comprises five basic units as follows:
1. Detector Unit
2. Master Indicator
3. Amplifier
4. The Gyro Unit
5. Control Panel

Detector unit

This is a very small unit in size and contains the sensing element. A simple flux valve as explained above would do as a sensing element but there are two disadvantages to its use. First, at varying latitudes, varying current will flow; second, an ambiguity exists. With relation to fig. 24.1 above we said that the flux on headings 030(M) and 150(M) will be of the same intensity but of opposite polarity. That's OK, then, we can differentiate. On heading 330(M), however, as the cosine of 330 is also positive, the flux will be of the same intensity and polarity as on heading 030(M).

These disadvantages are overcome by the use of three simple flux valves instead of just one. They are placed 120° apart and all six cores (or bars) are excited by a common primary coil. To improve sensitivity, six curved collector horns are attached to the outer ends of the bars. They collect the flux and feed into their respective spokes. (Fig. 24.7)

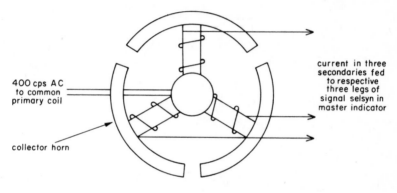

Fig. 24.7

Signal selsyn

This unit, located in the master indicator, receives the current from the detector unit. The selsyn unit consists of three legs similar to the spokes of the detector unit and a moveable coil, called the rotor, at the centre. The stators having received signals from the respective three secondary coils resolve the field produced in a final resultant direction. Detector unit and signal selsyn are shown in fig. 24.8.

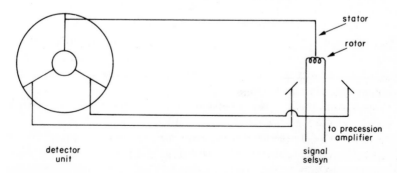

Fig. 24.8

Operation

Follow the operation with reference to fig. 24.9.

1. On switching on, the detector unit will produce voltages in three secondary coils. These voltages are a direct measure of the heading.

2. These voltages are fed to three respective stators of the signal selsyn. The stators resolve these voltages into a resultant field.

3. That resultant field is a definite direction. If the pointer of the master indicator is indicating that very heading at this time the rotor of the signal selsyn will be at 'null' position, that is, it will be lying at 90° to the direction of the resultant field. In this state no voltage is induced in the rotor.

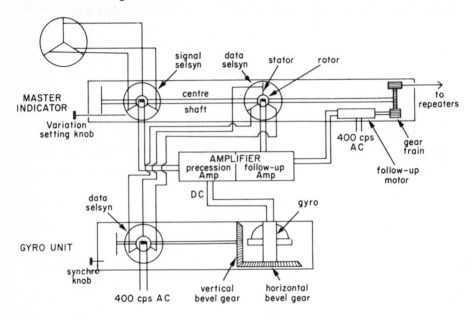

Fig. 24.9

4. If the heading is not synchronised, the rotor will not be at null and voltage will be induced in the rotor.

5. This voltage is fed to the precession amplifier where –
    a. it will be phase detected
    b. rectified (i.e. converted from AC to DC) and
    c. amplified

6. This DC signal is then passed to the precession coil in the Gyro Unit where the precession of the gyro takes place in the horizontal plane and in correct direction due to arrival of this signal. How this is done is explained later.

7. As gyro turns, the horizontal bevel gear turns, turning the vertical bevel gear with it. The vertical bevel gear carries a shaft on which are mounted the gyro unit pointer (this is the pilot's indicator) and the rotor of the gyro unit data selsyn. Therefore, as the vertical gear and the shaft rotate, the pointer and the rotor of the data selsyn must rotate too.

8. This data selsyn rotor of the gyro unit is continuously energised by AC and

therefore, has a standing field in it affecting its stators. When the rotor rotates, this field must rotate, producing new signals in its stators.

9.   These new signals are passed to the respective stators of the data selsyn in the master indicator, altering the field in them.

10. This new field in the stators will affect the rotor and an AC will be induced in it.

11. This AC or error signal is passed to the follow-up amplifier where it is amplified and then sent out to a two-phase follow-up motor.

12. This will energise the motor and it will start turning the central shaft in the MI (Master Indicator) via the gear train. Mounted on this shaft are the rotors of the two selsyns. They will continue to rotate until the rotor of the signal selsyn is at null position to the resultant field in that selsyn.

13. At this instant, signal selsyn rotor will stop sending error signals to the amplifier and to the rest of the system. The gyro is now aligned with the magnetic meridian, both the pointers are similarly synchronised with the meridian, and the system comes to rest.

Let's go over the whole thing again, this time taking some figures. Say, on switching on, the nose of the aircraft is pointing to 060(M). This is the field the detector unit will produce and the stators of the signal selsyn will resolve. Say, at this time the pointers of the two indicators are indicating 050(M).

Field in the stators − 060; pointers are not indicating this heading, therefore signal selsyn rotor is not at null. It sends off an error signal to the precession amplifier where three things noted above happen. The signal is then passed on to the precession mechanism of the gyro.

This turns the gyro, horizontal and vertical bevel gears. The shaft in the gyro unit rotates, rotating with it the rotor of the gyro unit data selsyn. This rotor was initially aligned with 050 indication. Now, say, at a given instant it has rotated to be aligned with 058 (in the course of its turn in the direction of 060). This turning will cause a new field in its stators which will be repeated in the stators of the data selsyn in the master indicator.

This rotor was aligned to indicate 050; now the field in its stators equals 058. A signal will be raised in the rotor, amplified and passed to the motor. The motor turns, turning the shaft and the rotors with it. Signal selsyn rotor turns to align with 058 − all rotors are in synchronism but the signal selsyn rotor has not gained the null position: it still has two degrees to go and therefore, it continues to send out error signals.

When all three rotors are aligned with 060, no more error signals are raised, and the compasses are synchronised.

Manual synchronising

In above illustration, the pointer will take exactly 5 minutes to move round from 050 to 060 since the rate of precession of the gyro is limited to 2° per minute. This slow rate of precession is quite adequate to keep the gyro in the magnetic meridian throughout the flight − i.e. check the gyro's apparent wander. But where a large discrepancy between the nose of the aircraft and the indicated heading is present, e.g. on initial switching on, the pointers are synchronised manually. This is done by use of the 'synchronising knob' on the Gyro Unit. (Fig. 24.10)

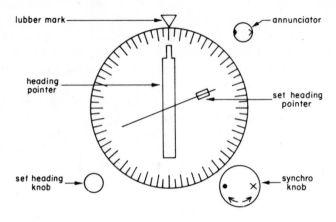

Fig. 24.10

The fact that the pointers are not indicating synchronised heading will be brought to your attention by the presence of a dot or a cross in the middle of the annunciator window on both the gyro unit indicator and the master indicator (which is the navigator's indicator). This indication is removed by depressing the synchro knob on the gyro unit and turning it in the direction shown by the arrow. When it is so depressed the gyro itself is disconnected and only the shaft and the pointers are turned.

Behaviour of CL2 in a turn

Earlier, we said that the normal precession operation is adequate to take care of the gyro wander etc. This might give an impression that the pointers would get desynchronised during a turn. This is not so, because during a turn the gyro, due to its rigidity, drives the pointers and keeps them aligned with the instantaneous heading. This is what takes place.

The aircraft, and therefore the gyro case, rotates about the horizontal bevel wheel and the gyro. Thus, the vertical bevel gear goes round the horizontal bevel gear, and the shaft in the gyro unit rotates. Error signals raised in the stators of the gyro unit data selsyn are repeated in the stators of the master indicator data selsyn. The rotor in that data selsyn raises an error signal, the motor turns and the shaft in the master indicator turns. This keeps the rotors of the signal selsyn and data selsyn in synchronism with the instantaneous heading during a turn.

During all this time the detector unit is also turning with the aircraft. Therefore the field in the signal selsyn stators is also rotating. Both this field and the rotor rotate at the same speed and in the same direction. Therefore, continuous synchronism is maintained.

In a steep and prolonged turn a slight desynchronisation may be expected. This is due to the fact that the detector unit is no longer horizontal and a small component of Z will enter in it, giving rise to a false field. However, on coming out of the turn, the needle will be brought to correct reading by the normal precession process. Apart from this small error there are no turn and acceleration errors.

Variation setting

The system incorporates a facility for setting variation on the compass so that true
headings may be flown. This is done by turning the stators of the signal selsyn which
has the effect of rotating the field in the stators by an equivalent amount. The facility
for setting variation is located on the master indicator and as the knob is turned a
scale in the variation window rotates behind a lubber line to indicate the amount of
variation being set.

Precession mechanism

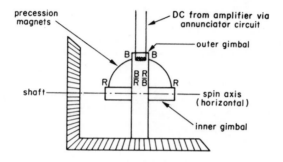

Fig. 24.11

The precession mechanism consists of a coil, called the precession coil, mounted
in the outer gimbal, and a pair of horn shaped permanent magnets mounted on the
inner gimbal. See fig. 24.11. The polarity of the permanent magnets is as shown in
the figure. When desynchronisation occurs, a DC is passed from the amplifier via the
annunciator circuit to the precession coil. (As a matter of interest AC models are also
available). Depending on the direction of the current, two ends of the precession coil
will acquire magnetic polarity. At one end this polarity will be in attraction and at
the other end in repulsion to the polarity of the permanent magnet. This attraction
and repulsion will apply a torque on the inner gimbal in the vertical plane and the
gyro will precess in the horizontal plane, that is, the horizontal bevel gear will turn.

Erection mechanism

It will be appreciated that, as with the DGI, the gyro in CL2 must be maintained
horizontal. Erection mechanism ensures this. It consists of a two-phase torque motor
with its stators mounted on the outer gimbal and a levelling switch mounted on the
inner gimbal — see fig. 24.12.

The levelling switch is made up of a commutator split in two segments by an
insulating strip. When the gyro axis is horizontal the stators rest on the insulated
strip and the switch to the motor is off — fig. 24.12(a). When the axis is tilted the
stator will come in contact with the commutator and the circuit to the torque
motor is complete. The direction of the current to the torque will depend on the
direction of the tilt, as is seen in fig. 24.12(b). The direction of the current decides
the direction the torque will be applied, that is, clockwise or anti-clockwise. The
torque so applied about the vertical axis will precess the gyro axis until it is once

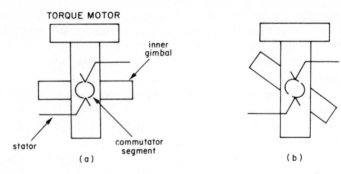

Fig. 24.12

again horizontal when the stator will slip off the commutator and rest on the insulated strip, switching off the motor.

The annunciator

This unit gives the indication of synchronisation state, and consists of a pivoted arm carrying a flag marked with dot and cross at one end and a permanent magnet at the other. The magnet is held between two annunciator coils connected in series. When the pointers are desynchronised from the magnetic meridian DC flows from pre-

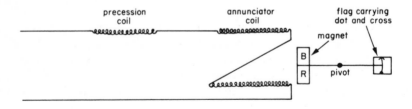

Fig. 24.13

cession amplifier to the precession coil. Use is made of this current flow which passes through the annunciator circuit. As it passes through the two annunciator coils (connected in series) the two ends nearest to the permanent magnet will acquire opposite polarity, one end attracting the magnet and the other end repelling. As the magnet is drawn towards one end of the coil the flag arm will turn about its pivot, the flag will be displaced from the centre and one of its ends, carrying either a dot or a cross, will appear in the window. Whether the dot or a cross will appear in the centre of the window depends on which end of the magnet is drawn towards the coils, which in its turn depends on the direction of the DC.

Control panel

With CL2 system there are two gyro units, one for each pilot, together with the master indicator installed at the navigator's position. But as we saw above only one gyro unit could be run together with the detector unit. In fig. 24.9 we have no accommodation for two gyro units. Therefore, the other gyro must run as a DGI. On the control panel there is a three-position switch on which you may select gyro off, gyro on port, or gyro on the starboard side. In the off position, both gyros operate as a DGI, that is, unmonitored by the detector unit.

Adjustment and calibration

Having done a normal swing and calculated coefficients A, B and C, proceed as follows to correct for them.

Equipment Required. A centre reading voltmeter (CRV). This is plugged in at the jack socket in the amplifier. This then by-passes the annunciator circuit; the CRV acts as a more sensitive annunciator. If no current flows through the CRV the system is synchronised. You also require a stable DC supply of 24 – 28V and a Compass Corrector Key.

Correction for B and C.

1.   Having worked out the heading you want the compass to read, press the manual synchro knob on the gyro unit, turn it and give the required reading to the pointer.
2.   Keep the knob depressed. If the knob is released the pointer will be precessed back to the original reading.
3.   The CRV needle is now displaced from its zero position to one side or the other.
4.   On the corrector box there is a two-position switch. In one position corrections up to $3°$ may be made; in the other, corrections over $3°$ may be made. If your correction does not exceed $3°$ place the switch to $3°$ position, otherwise on to $15°$ position.
5.   Insert the compass corrector key in B or C socket as appropriate and turn the key until CRV reads 0.
6.   Correction is complete; release the synchro knob.

Coefficient A over $2°$

1.   Steps 1, 2 and 3 are as per B and C above.
2.   Turn the detector unit physically until CRV reads 0.
3.   Release the synchro knob.

Coefficient A – $2°$ and below

This is simply taken out on the Variation scale on the master indicator as follows:

1.   Set the value of the coefficient A on the VSC as if you are setting variation. If A is positive set easterly variation of equivalent amount. If it is negative set westerly variation. For example you wish to correct for A, value -1. Set $1°$ westerly variation.
2.   Insert compass corrector key in the groove on the side of the master indicator (approximately at the 8 o'clock position) and turn the key until the lubber line of the VSC moves and covers 0 variation. The lubber line was displaced by you when you carried out step 1 above.

Advantages of a gyrosyn compass

The advantages of a gyrosyn compass over a DGI are these:

1.   DGI suffers from slow drift and has to be reset frequently in a flight. Further, when resetting, the aircraft must be flown straight and level to be able to take reading of a magnetic compass. CL2 does away with this, since its indicators are monitored by a detector unit.
2.   The detector unit can be installed in a remote part of the aircraft where influences due to aircraft magnetism are least. This gives a better accuracy to CL2 whereas a

magnetic compass from which the DGI is reset is exposed to these influences.

3.   The flux valve used in the detector unit 'senses' the meridian instead of 'seeking' it. This avoids northerly turning errors.

4.   The compass can be detached from the detector unit by a flick of a switch and the gyro then works as normal DGI. This entirely does away with the requirement of a pure DGI in the cockpit. (CL2 is used as a DGI when flying in extreme northerly/ southerly latitudes where H is very feeble.)

Section 4

SOLAR SYSTEM

25: Solar System: Time

The Solar System consists of the Sun, nine major planets of which the Earth is one, and about 2 000 minor planets or asteroids. All members of the solar system are controlled by the Sun which is distinguished by its immense size and its radiation of light and heat; for all practical purposes, it may be considered as the stationary centre round which all the planets revolve.

Unlike the Sun, the planets and their satellites are not self-luminous, but reveal their presence by reflecting the Sun's light. The planets revolve about the Sun in elliptical orbits, each one taking a period of time about the job: Mercury takes 88 days, for example, while Pluto which is rather a long way from the parent body, is thought to take about 248 years. The planetary satellites in the meantime are revolving about their own parents.

Certain laws relating to the motion of planets in their orbits were evolved by the astronomer Kepler, who died in abject poverty as a reward:

 (i) each planet moves in an ellipse, with the Sun at one end of its foci;
 (ii) the radius vector of any planet sweeps out equal areas in equal intervals of
      time.

These are the important laws for our purpose in studying the Earth's motion, as we shall see.

The Earth rotates on its axis in a West to East direction, resulting in day and night. It revolves round the Sun along a path or orbit which is inclined to the Earth's axis at about $66\frac{1}{2}°$, resulting in the seasons of the year. When the Earth is inclined towards the Sun, we get the Summer Solstice (June 21); when the axis is away from the Sun, we get the Winter Solstice (Dec 22). When the Earth's axis is at right angles to the

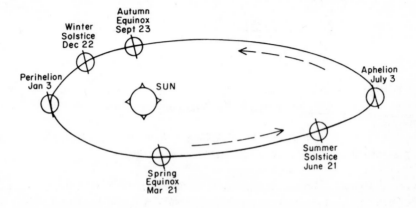

Fig. 25.1

Sun, days and nights are equal, the Spring and Autumn Equinox (March 21 and Sept 23). The point where the planet is nearest to the Sun is called perihelion, and where farthest aphelion; it is worth noting that in obeying Kepler's second law, the speed of the Earth at perihelion is faster along its orbit than at aphelion; and that the Earth is nearer to the Sun in British winter.

Measurement of time

The instant at which a heavenly body is directly over a meridian is called a 'transit'. The earth rotates on its axis from West to East and to us on its crust, the heavenly bodies appear to revolve about the Earth from East to West. The period of this apparent revolution is measured by the time elapsing between two successive transits of a heavenly body, called a 'day'. If a star were the heavenly body, since stars are such immense distances away, the 'sidereal' day would be of constant length, for only the Earth's axial movement of rotation would affect the star's apparent movement. The Earth is 8 light minutes from the Sun, but $4\frac{1}{3}$ light years from the nearest star. We measure the day by the Sun, of course, and must needs investigate the problems that result, without involving ourselves with the vast academic matters of the Universe.

An 'apparent solar day' is the time interval between two successive transits of the real or apparent Sun at the same meridian, the word 'apparent' signifying that is how it appears to us as we move with the Earth. Seen from above the North Pole, the Earth rotates on its axis at an even speed in an anti-clockwise direction, and also revolves round the Sun in its orbit in an anti-clockwise direction; since the direction of axial rotation is the same as orbital revolution, West to East, it follows that it must rotate on its axis through more than 360° to produce successive transits.

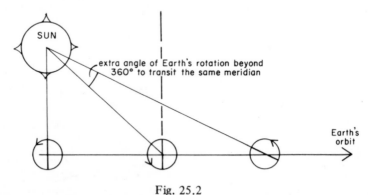

Fig. 25.2

Furthermore, since the Earth's orbit is elliptical, the Earth's speed of revolution on its orbit varies continuously and the length of an Apparent Solar Day will not be constant.

The time of transit is called Local Apparent Noon; but these facts and their unequal variations would make human working life difficult.

Mean Solar Time

In order to have a constant measurement of time which will still have the Solar Day as a basis, the average length of an apparent solar day is taken, called the Mean Solar Day, divided into 24 hours of mean solar time. This is arrived at as follows:

The time of orbital revolution of the Earth in one year is constant at 365 days 5 hours 48 minutes 49 seconds. A mean sun is imagined to travel once round a circular path in the same plane as the Equator at a constant speed in the same time as the True Sun travels round its apparent elliptical path at its varying speeds: thus, the time of two successive transits of the Mean Sun over any given meridian is constant, as both the Earth and Mean Sun are rotating in the same plane at constant speeds. In other words, we keep time by an imaginary or Mean Sun which leaves a meridian and returns to it 24 hours later of mean time.

The discrepancy between the transit of the Apparent Sun and the Mean Sun over the Greenwich meridian is not large enough to affect the working day (Mean noon at Greenwich is approximately $16\frac{1}{3}$ minutes later than Apparent noon in November; and $14\frac{1}{3}$ minutes earlier in February).

The year of 365 days 5 hours 48 minutes 49 seconds (again measuring in Mean Solar Time) is itself an inconvenient measure: the Calendar year is 365 days, so to even things up, a leap year of 366 days is inserted every 4 years, and to round things off, 3 leap years are suppressed every 4 centuries.

Time in arc

Time can also be measured in arc since in one day of mean solar time, the Sun is imagined to travel in a complete circle round the Earth, a motion of 360°. The measurement of 24 hours then is the same measurement as 360° of longitude. Really, that is the only thing to remember, for any conversion of one to the other is mental arithmetic; for reference:

$$24 \text{ hours} = 360° \text{ of longitude}$$
$$1 \text{ hour} = 15° \text{ of longitude}$$
$$1 \text{ minute} = 15' \text{ of longitude}$$
$$1 \text{ second} = 15'' \text{ of longitude}$$

Conversely:

$$360° = 24 \text{ hours of time}$$
$$1° = 4 \text{ minutes of time}$$
$$1' = 4 \text{ seconds of time}$$
$$1'' = 1/15 \text{ second of time}$$

Local Mean Time

The beginning of the day at any place is midnight, or 0000 hours LMT, e.g. 0000 hours at a place in longitude 50W will be when the Mean Sun is in transit with the ante meridian 130°E. This means that the LMT of places in different longitudes varies by an amount corresponding to the change in longitude. A standard meridian has to be fixed to which all LMTs can be referred, and the meridian at Greenwich is internationally accepted as this standard. Local Mean Time at Greenwich is called Greenwich Mean Time.

The Greenwich Day commences when the Mean Sun is in transit with the ante meridian of Greenwich and as the Sun appears to travel from East to West, it will be in transit with Easterly meridians before it is in transit with Greenwich. Thus, LMT of places East of Greenwich will be ahead of GMT and places West of Greenwich will be behind GMT.

Consider these diagrams for a Sunrise at 0600 hrs LMT, in order to establish the relationship of LMT to time on the Greenwich meridian.

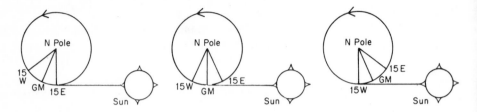

Fig. 25.3

The rule is:

Longitude East, Greenwich Time Least

Longitude West, Greenwich Time Best

All flying is done on GMT, no matter whereabouts on the globe a pilot is operating, but the Sun rises and sets at different times on the same meridian, depending on the celestial latitude (declination) of the Sun and the Earthly latitude of the observer; more later on this, but the times of the phenomena are listed in the Air Almanac in LMT, so the conversion from one to the other is very necessary knowledge indeed. A few examples:

(i) GMT 1200 hrs,   then LMT at 60°E = 1600 hrs

and LMT at 80°W = 0640 hrs

(ii) What is the GMT and GD of a place in longitude 46°25′E where LMT is 14h 23m 15s and LD is 2nd June?

46°25′ in time =   3h 05m 40s

LMT and LD = 14h 23m 15s   2nd June

GMT least   = 11h 17m 35s   and GD still

2nd June

(iii) If GD and GMT are 2nd Dec 21h 02m 24s, give LD and LMT in longitude 96°47′E.

96°47′ in time = 6h 27m 08s

GMT least   = 21h 02m 24s 2nd Dec GD

LMT   = 03h 29m 32s 3rd Dec LD

The Air Almanac, on one of the last odd pages of the book gives a table for the conversion of arc to time from 1° to 360°, so that 341° for example is at once read off as 22h 44m; the end column gives minutes of arc into minutes and seconds of time e.g. 17′ of arc is 1m 08s of time.

Standard Time

Standard Time is the set time used for a particular country or part of a country. In general, it is based on the LMT 7½° on either side of a regular meridian divisible by 15°, but this of course often conflicts with political and national requirements, and has finally become arbitrary. The U.K. normally kept GMT as its Standard Time, but now has added one hour to GMT. A country like the U.S.A. is too longitudinally vast to keep one standard time, and divides itself up, so that the Atlantic Seaboard keeps Eastern Standard Time, 5 hours behind GMT, while Hawaii is 10 hours slow on GMT.

All Standard Times are listed in the Air Almanac, in three tables: (1) Places East of Greenwich, fast on GMT, (2) Places on GMT, (3) Places West of Greenwich, slow on GMT. With many local adjustments and diverse boundaries of separation, there is a plenitude of footnotes to be watched: a pleasant note is that the adjustment one to the other of GMT to ST is explained in each table, to the chagrin of Ministry Examiners.

Dateline

When travelling Westward from Greenwich, an observer would eventually arrive at longitude 17959W, where the LMT is about to become 12 hours less than GMT. An observer travelling Eastward from Greenwich would eventually arrive at 17959E where the LMT is about to become 12 hours more than GMT. Thus there is a full day of 24 hours difference between the two travellers, although they are both about to cross the same meridian. When the ante-meridian of Greenwich is crossed, one day is gained or lost, depending on the direction of travel: the Dateline is the actual line where the change is made, and is mainly the 180 meridian, with some slight divergences to accommodate certain groups of South Sea Islands and regions of Eastern Siberia. The problem readily resolves itself in flying – your watch is always on GMT: the place whose Standard Time you want is listed in the Air Almanac: apply the correction to GMT, and the date will take care of itself. Crossing the dateline always seems to excite passengers in aircraft much more than crossing the Equator: you will want to let them know when, on a leg from HONOLULU to TOKYO, the date changes from the 2nd to the 3rd.

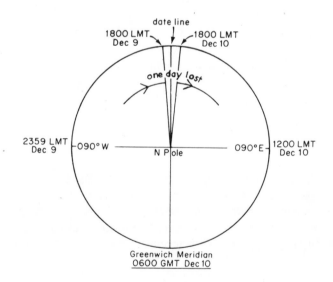

Fig. 25.4

Sunrise and sunset

We are only concerned with the visible phenomena, when the centre of the Sun is coincident with the observer's horizon: at the moment of this vision, the centre of the Sun is actually $1°$ below the horizon, but due to refraction we see it as higher than it really is. All this is mercifully built into the Tables to let the pilot cope with what he sees, and let the theory look after itself.

The times of S/R and S/S at any place change by only a minute or two each day except in high latitudes; so the time of the occurrences at specified Latitudes on the

Greenwich meridian may be taken as the same for all longitudes. A tabulation of 0708 for a certain date and latitude is in fact the LMT of the phenomenon on the Greenwich meridian, it will also be the LMT on any other longitude on the same local date and Latitude: by the time 0708 LMT comes along at 30°W, the time on the Greenwich meridian is by now 0908. And if you are at 30°W in an aeroplane with your watch set to GMT as usual, you will look up the Air Almanac for your Latitude, find 0708 LMT and apply the usual rule to find the GMT at the time of the rising. Check again with fig. 25.3.

The tabulation in the Air Almanac covers every band of Latitude for dates three days apart: the bands of Latitude are sometimes 10° apart, sometimes 2°. The times of S/R and S/S vary considerably as the Latitude of the observer increases. The interpolation from the tables for the date required on the observer's Latitude may be done without tedious calculation: take the nearest tabulated day, and interpolate on that day for the Latitude required.

The diagrams which follow will explain the effect of Latitude; and, bearing in mind that the Sun's declination is a maximum $23\frac{1}{2}$°N or $23\frac{1}{2}$°S approximately, how it comes about that the Sun can be above or below the horizon all day in some Latitudes. CH is the observer's horizon.

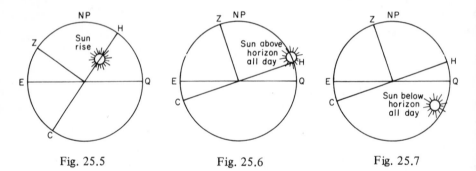

Fig. 25.5   Fig. 25.6   Fig. 25.7

With a midsummer's day N Hemisphere, the length of the day from S/R to S/S from fig. 25.5 increases with the observer's latitude Z until eventually, in fig. 25.6, he is Northerly enough to have the Sun above his horizon all day: this particular observer gets his come-uppance as the Sun makes its way South, for then it is continuous Northern night (fig. 25.7). The symbols in the Tables for such occasions are explained at the beginning of the Air Almanac, after the daily pages.

Sample: What is GMT of S/S in Position 32°00′S 172°15′W on 13th Jan LD?
The extract from the Air Almanac looks like this:

	Jan 11	Jan 14
Lat. S30°	1906	1905
Lat. S35°	1918	1917

and 13th Jan is thus as for 14th Jan. Interpolate for 32°S, 2/5 of 12 minutes = 5 minutes. Thus

$$
\begin{array}{rl}
\text{13th Jan at } 32°00'S = & 1910 \text{ LMT} \\
172°15'W \text{ to time} & \underline{1129} \\
\text{GMT} & 3039 \quad \text{best} \\
= & 0639 \quad \text{14th Jan GMT \& GD}
\end{array}
$$

i.e. the time and date on the Greenwich meridian when the sun sets in 32°00'S
172°15'W.

Twilight

When the Sun is below the horizon, an observer will still receive light which has been
reflected and scattered by the atmosphere. It is divided into three stages: Astronomi-
cal (12° below Horizon, and it's dark); Nautical (6° – 12° below the Horizon, and
has something to do with the sea horizon being indistinct, and artificial light being
required to box the compass abaft the foc'sle); and Civil Twilight when the Sun's
centre is actually between 1° and 6° below the horizon, when work is possible
without artificial light, and the stars are not clearly visible. This last is the one we're
concerned with.

The Air Almanac tabulates in exactly the same form as S/R and S/S the beginning
of morning twilight and the end of evening twilight, all in LMT as before. The
duration of morning twilight is then simply the difference between S/R and the
twilight tabulation for that place on that day. Just as in high latitudes we had
occasions of 24 hours daylight or darkness, so again we shall have occasions where
twilight lasts all night. The symbol is explained in the same place as before in the Air
Almanac.

The whole period of twilight has particular significance to pilots beyond the
obvious transition of light and dark, and the impairment of visual judgments at that
time. It is around twilight that the ionosphere starts to move, with the consequent
reduction of the effective range of MF radio aids especially, and of HF aids to some
extent. The radio compass becomes sluggish; a cathode ray tube presentation in an
aircraft working a station well within its normal ground wave range gets thoroughly
grassed up; even HF R/T suffers increased noise with the altering properties of the
atmosphere for radio transmission as the Sun's powers change. The navigator is
meanwhile denied the heavens for astro-navigaton. It is worth a thought that in fast
aircraft and long range radio aids, a twilight zone can exist between the aircraft and
the station for quite a long time, most noticeable on Easterly and Westerly flights.

Effect of height

The tabulations in the Air Almanac for risings and settings and twilight are all for
an observer at sea level; in flight, the visible horizon is extended, and the phenomena
will occur earlier in the morning and later in the evening. The rate of movement of
the Sun through the angle of depression varies with the latitude of the observer and
the time of year as well as with the observer's height, and the calculation of the time
difference, though measurable, takes the pilot into matters at present outside his
syllabus of study. An explanation and all the necessary figures are in the Air Almanac.

The Moon

The Moon is a perfect sphere which travels round the Earth according to Kepler's
Laws 1 and 2, but due to the Earth's own orbit its apparent path is somewhat
irregular though calculable. It rotates on its axis West to East in exactly the same
time as it takes to get round the Earth, so that we are always presented with the
same face (actually, over a period, 59% of the Moon's surface has been seen by the
Earthbound observer). One half of the Moon is always illuminated by the Sun, and

as the Moon revolves round the Earth, the sunlit half is presented in varying amounts, the phases of the Moon.

When the Moon passes between the Earth and the Sun, it is said to be in conjunction, and its sunlit half is away from the Earth, a new Moon. It is in opposition when the Earth lies between the Sun and Moon, the illuminated face is straight towards us, a full Moon. When the direction of the Moon is 90° to that of the Sun, it is in quadrature, when half of the disc is visible from the Earth, called the first and last quarter; crescent phases are less than half full, gibbous more than half. The age of the Moon is the number of days which have elapsed since it was last new, and even that is recorded in the daily pages of the Air Almanac.

In fig. 25.8 which illustrates the phases, it must be borne in mind that the three bodies are not in a straight line; if they were, then an eclipse, total or partial, solar or lunar, would be seen from the Earth. In simple terms, a latitude projected into the celestial sphere is called the declination of a heavenly body, and the declinations of the Sun and Moon must be nearly equal to give a solar eclipse; and there must be a relationship for a lunar eclipse when the Moon passes through the Earth's shadow.

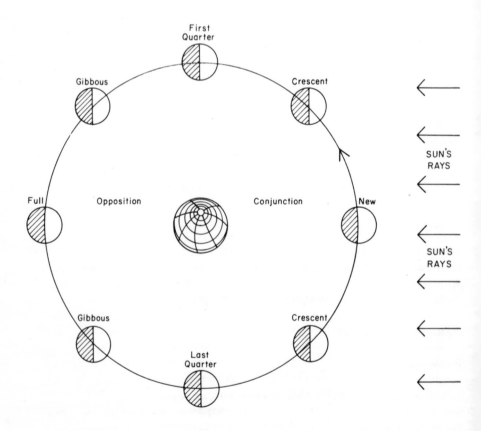

Fig. 25.8

Moonrise and moonset

The Moon revolves on its orbit West to East round the Earth once in $29\frac{1}{2}$ days: its average daily movement =

$$\frac{360}{29\frac{1}{2}} = 12\frac{3}{4}° \text{ which is 51 minutes of time.}$$

If the declination of the Moon (i.e. its celestial latitude) were constant, it would rise and set on the average 51 minutes later each day.

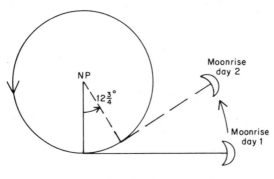

Fig. 25.9

But the declination changes very rapidly, and the difference between M/R or M/S on successive days is marked and inconstant. The extra amount of revolution required to transit the same meridian successively is known as the DAILY LAG, and it varies with the latitude of the Observer and the declination of the Moon. So the daily changes of the M/R and M/S are great enough to prevent figures tabulated in the Air Almanac for the phenomenon on the Greenwich Meridian to be regarded as LMTs for other meridians.

Precise times of M/R and M/S are seldom required however: and a tabulation of two different daily lags for correction to Easterly or Westerly longitude is avoided in the Almanac by inserting under the term 'Difference' half the daily lag. This 'difference' is found against the observer's latitude, on the daily pages, with the LMT of M/R of M/S. This difference is then taken to the 'Interpolation' of Moonrise/Moonset (on the loose flap usually), and entering with difference and observer's longitude, the correction is read off: this correction is added to LMT M/R or M/S if Longitude W, subtracted if Longitude E. A thumbnail sketch would illustrate why – the lag is progressive, and East of Greenwich the phenomenon will occur at an earlier LMT than at Greenwich, so the correction is subtracted: on Westerly meridians, the correction is additive since the Moon will have moved further on its orbit.

As the time interval between successive phenomena is more than 24 hours, it follows that in each month there will be one day when there will be no M/R and another when there will·be no M/S.

This will occur when M/R or M/S is near to local midnight. The tabulations are for the phenomena on the Greenwich meridian; in the illustration above, it is clear that the effect of the lag for M/S on Easterly meridians may not have caused the jump from local Friday to local Sunday, i.e. the M/S could have occurred on late local Saturday. To allow for this, the Air Almanac gives the time as 2400+, as a basis for

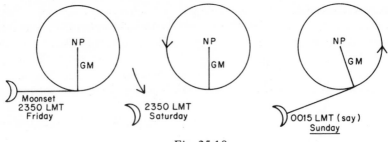

Fig. 25.10

calculation of the occurrence in other longitudes; the point to watch is the date. 2431 as the tabulated time on 7th Jan is really 0031 on 8th Jan, the date at Greenwich.

It can happen that the times found are for the day before, or day after, the one required: in which case, it is quite in order to add or subtract twice the difference, watching the date carefully, as:

Required GMT of M/R 8th March 1968, at 52°N, 150°E.

From daily Page M/R 8/3/68	0958	diff. 25
From Flap Table 25 v 150	− 22	
LMT 8/3/68	0936	
Long. 150°E	1000	
GMT	2336	Least 7/3/68
Twice diff.	+ 50	
GMT + one day	0026	9th March

In other words, there was no M/R in this position on 8th March, GD.

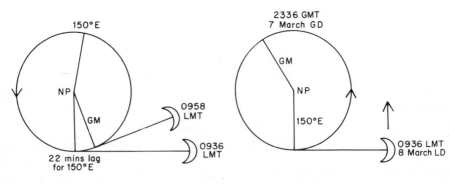

Fig. 25.11   Fig. 25.12

Fig. 25.11 is a sketch of the tabulated M/R for the Greenwich meridian with times at the occurrence for Greenwich meridian and 150°E. Knowing then that the M/R takes place (fig. 25.12), 10 hours (approx.) earlier for 150°E than on the Greenwich meridian, the Greenwich time will be 2336, 7th March 1968 GD and no moonrise will take place at 150°E on the 8th, for the lag (or double the difference) will whip the next occurrence there to the 9th March, *the calculations being all for GD.*

A final touch about the tabulations of M/R and M/S on the daily pages. With this wretched body so near the Earth, there is plenty of parallax and refraction, all allowed for in the Tables, so that the usual rule of visible rising and setting is obeyed: the differences in higher latitudes are often too erratic to be reasonably measured, and an asterisk is inserted to warn you. If the difference is noted as negative the correction duly found from the Interpolation table on the flap is applied in the opposite direction, as the footnote points out. On this table, too, it is accurate enough to take the nearest noted figure for difference v longitude without interpolation.

It's a lot of hard work for very little value when flying. A full moon at night can make for less tedious flying, I suppose, but a navigator would never use the thing if the stars are visible, and it hardly ever interferes with his sights; though by day, in the right position in the celestial globe, it can provide a Sun/Moon Fix, but that's the navigator's problem.

Practice problems

Use the tables at the end of this chapter to extract the times of sunrise, sunset and twilight. You will need to interpolate between the dates and the latitude bands. The answers should be correct to the nearest minute.

1. If the LMT in position $47°30'N\ 00°00'$ is 12 00 hr what is the LMT in the position $54°20'S\ 00°00'$?

Answer: 12 00 hr.

2. LMT in position $54°40'S\ 00°00'$ is 15 25 hr, what is the GMT in position $15°S\ 180°$?

Answer: 15 25 hr.

3. If the GMT is 10 27 30 hr what is the LMT at $47°32'W$?

Answer: 07 17 22 hr LMT.

4. At 08 00 hr GMT, what is the LMT at $40°30'N\ 117°15'E$?

Answer: 15 49 00 LMT.

5. What is the LMT in Istanbul ($41°N\ 29°E$) when the
   (a) LMT in Paris (long $02°30'E$) is 08 00 hr?
   (b) LMT in Rome (long $12°E$) is 19 00 hr?
   (c) LMT in Baghdad (long $44°17'E$) is 00 13 00 hr on 1st May?

Answers: (a) 09 46 00 hr LMT; (b) 20 08 00 hr LMT; (c) 23 11 52 hr LMT on 30th April.

6. The LMT at Hamilton, Ontario ($43°09'N\ 79°56'W$) is 21 16 00 hr on 17th June. What is the (a) GMT? (b) ST (standard time)? (Note: Hamilton ST is 5 hr behind GMT.)

Answers: (a) 02 35 44 hr 18th June; (b) 21 35 44 hr 17th June.

7. The LMT at Gibraltar ($36°07'N\ 05°21'W$) is 11 15 hr on 4th July. What is the (a) GMT? (b) ST? (Gibraltar ST is 1 hr ahead of GMT.)

Answers: (a) 11 36 24 hr on 4th July; (b) 12 36 24 hr on 4th July.

8. The GMT at El Aovina, Tunisia ($36°50'N\ 10°14'E$) is 19 47 hr on 28th April. What is the (a) LMT? (b) ST? (Tunisia ST is 1 hr ahead of GMT.)

Answers: (a) 20 27 56 hr on 28th April; (b) 20 47 00 hr on 28th April.

9. The standard time in Damascus, Syria ($33°29'N\ 36°14'E$) is 21 59 00 hr on 5th August. What is the (a) GMT? (b) LMT? (Syria ST is 2 hr ahead of GMT.)

Answers: (a) 19 59 00 hr on 5th August; (b) 22 23 56 hr on 5th August.

10. An aircraft is making for Rawalpindi, West Pakistan ($32°25'N\ 76°20'E$) on

9th February 1968, and pinpoints over Lahore at 11 42 GMT. If the distance to go is 173 nm and the W/V is L/V, what RAS must be maintained at FL 60, temp. +30°C to make the destination by sunset?

Answer: RAS 176 kt (computer). (Note: Rawalpindi sunset time 17 39 37 hr.)

11. If the GMT is 12 00 hr, in what longitude is the LMT 20 00 hr?

Answer: 120°E.

12. Find the ST of sunset at Leningrad, Russia (59°46'N 30°20'E) on local date 22nd January 1968. (Russia, west of long 40°E, ST is 3 hr ahead of GMT.)

Answer: 16 46 26 hr ST on 22nd January.

13. Give the GMT of sunrise at Auckland, New Zealand (36°50'S 174°48'E) on local date 3rd February 1968.

Answer: 17 37 46 hr GMT on 2nd February.

14. Give the GMT sunset at Taranto (40°26'N 17°16'E) on local date 6th February 1968.

Answer: 16 13 44 hr GMT 6th February.

15. An aircraft wishes to arrive at Sydney, NSW (33°52'S 151°23'E) one hour before sunset on 9th January 1968. The flight is planned to take 5 hr 13 min. What is the latest time LMT it can depart from 35°S 172°17'E?

Answer: 14 25 52 hr LMT on 9th January. (Sydney sunset LMT 19 15 16 hr.)

A62

SUNRISE

Lat.	Dec. 30	Jan 2	Jan 5	Jan 8	Jan 11	Jan 14	Jan 17	Jan 20	Jan 23	Jan 26	Jan 29	Feb 1	Feb 4	Feb 7	Feb 10	Lat.
	h m	h m	h m	h m	h m	h m	h m	h m	h m	h m	h m	h m	h m	h m	h m	
N 72	■	■	■	■	■	■	■	■	■	■	11 06	10 35	10 11	09 49	09 30	N 72
70	■	■	■	■	■	■	■	11 14	10 48	10 27	10 08	09 51	09 34	09 18	09 03	70
68	■	■	11 37	11 15	10 57	10 42	10 28	10 14	10 01	09 47	09 34	09 21	09 09	08 56	08 43	68
66	10 32	10 27	10 22	10 15	10 07	09 59	09 50	09 40	09 30	09 20	09 10	09 00	08 49	08 38	08 27	66
64	09 52	09 49	09 46	09 41	09 36	09 30	09 23	09 16	09 08	09 00	08 51	08 42	08 33	08 24	08 14	64
62	09 24	09 23	09 20	09 17	09 13	09 08	09 03	08 57	08 50	08 43	08 36	08 28	08 20	08 12	08 03	62
N 60	09 03	09 02	09 00	08 58	08 55	08 51	08 46	08 41	08 35	08 29	08 23	08 16	08 09	08 01	07 54	N 60
58	08 46	08 45	08 44	08 42	08 39	08 36	08 32	08 28	08 23	08 18	08 12	08 06	07 59	07 53	07 46	58
56	08 32	08 31	08 30	08 28	08 26	08 23	08 20	08 16	08 12	08 07	08 02	07 57	07 51	07 45	07 39	56
54	08 19	08 19	08 18	08 17	08 15	08 12	08 10	08 06	08 02	07 58	07 54	07 49	07 43	07 38	07 32	54
52	08 08	08 08	08 08	08 06	08 05	08 03	08 00	07 57	07 54	07 50	07 46	07 42	07 37	07 32	07 26	52
N 50	07 58	07 58	07 58	07 57	07 56	07 54	07 52	07 49	07 46	07 43	07 39	07 35	07 30	07 26	07 21	N 50
45	07 38	07 38	07 38	07 38	07 37	07 36	07 34	07 32	07 30	07 27	07 24	07 21	07 18	07 14	07 10	45
40	07 21	07 22	07 22	07 22	07 22	07 21	07 20	07 18	07 16	07 14	07 12	07 09	07 07	07 03	07 00	40
35	07 07	07 08	07 08	07 08	07 08	07 08	07 07	07 06	07 05	07 03	07 02	07 00	06 57	06 55	06 52	35
30	06 55	06 56	06 57	06 57	06 57	06 57	06 57	06 56	06 55	06 54	06 53	06 51	06 49	06 47	06 45	30
N 20	06 34	06 35	06 36	06 37	06 37	06 38	06 38	06 38	06 38	06 37	06 37	06 36	06 35	06 34	06 32	N 20
N 10	06 16	06 17	06 18	06 19	06 20	06 21	06 22	06 22	06 22	06 23	06 23	06 23	06 22	06 22	06 21	N 10
0	05 59	06 00	06 01	06 03	06 04	06 05	06 06	06 07	06 08	06 09	06 09	06 10	06 10	06 11	06 11	0
S 10	05 41	05 43	05 44	05 46	05 47	05 49	05 51	05 52	05 54	05 55	05 56	05 57	05 58	05 59	06 00	S 10
20	05 22	05 24	05 26	05 28	05 30	05 32	05 34	05 36	05 38	05 40	05 42	05 43	05 45	05 47	05 49	20
S 30	05 00	05 02	05 05	05 07	05 09	05 12	05 14	05 17	05 20	05 23	05 25	05 28	05 30	05 33	05 35	S 30
35	04 48	04 50	04 52	04 55	04 57	05 00	05 03	05 06	05 09	05 12	05 15	05 19	05 22	05 25	05 28	35
40	04 33	04 35	04 38	04 40	04 43	04 47	04 50	04 53	04 57	05 01	05 04	05 08	05 11	05 15	05 19	40
45	04 15	04 17	04 20	04 23	04 27	04 30	04 34	04 38	04 42	04 47	04 51	04 55	05 00	05 04	05 09	45
50	03 53	03 55	03 59	04 02	04 06	04 10	04 15	04 20	04 25	04 30	04 35	04 40	04 45	04 51	04 56	50
S 52	03 42	03 45	03 48	03 52	03 56	04 01	04 06	04 11	04 16	04 22	04 27	04 33	04 38	04 44	04 50	S 52
54	03 30	03 33	03 37	03 41	03 45	03 50	03 55	04 01	04 07	04 13	04 19	04 25	04 31	04 37	04 43	54
56	03 16	03 19	03 23	03 28	03 32	03 38	03 44	03 50	03 56	04 03	04 09	04 16	04 23	04 29	04 36	56
58	02 59	03 03	03 07	03 12	03 18	03 24	03 30	03 37	03 44	03 51	03 58	04 06	04 13	04 20	04 28	58
S 60	02 39	02 43	02 48	02 54	03 00	03 07	03 14	03 21	03 29	03 37	03 45	03 54	04 02	04 10	04 19	S 60

A62

SUNSET

Lat.	Dec.	January										February				Lat.
	30	2	5	8	11	14	17	20	23	26	29	1	4	7	10	
°	h m	h m	h m	h m	h m	h m	h m	h m	h m	h m	h m	h m	h m	h m	h m	°
N72	■	■	■	■	■	■	■	13 09	13 37	13 59	13 21	13 53	14 18	14 40	15 00	N72
70	■	■	■	■	■	■	■	14 08	14 24	14 38	14 19	14 38	14 55	15 11	15 27	70
68	■	■	12 34	12 59	13 18	13 36	13 52	14 42	14 54	15 05	14 53	15 07	15 20	15 34	15 47	68
66	13 33	13 40	13 49	13 58	14 09	14 19	14 31	15 06	15 16	15 26	15 17	15 28	15 40	15 51	16 02	66
64	14 13	14 18	14 25	14 32	14 40	14 48	14 57	15 26	15 34	15 42	15 36	15 46	15 56	16 05	16 15	64
62	14 40	14 45	14 50	14 56	15 03	15 10	15 18	15 41	15 49	15 56	15 51	16 00	16 09	16 17	16 26	62
N60	15 01	15 05	15 10	15 15	15 21	15 28	15 34	15 41	15 49	15 56	16 04	16 12	16 20	16 28	16 35	N60
58	15 19	15 22	15 26	15 31	15 36	15 42	15 48	15 54	16 01	16 08	16 15	16 22	16 29	16 36	16 44	58
56	15 33	15 36	15 40	15 45	15 49	15 55	16 00	16 06	16 12	16 18	16 25	16 31	16 37	16 44	16 51	56
54	15 46	15 49	15 52	15 56	16 01	16 06	16 11	16 16	16 22	16 27	16 33	16 39	16 45	16 51	16 57	54
52	15 56	15 59	16 03	16 07	16 11	16 15	16 20	16 25	16 30	16 35	16 41	16 46	16 52	16 57	17 03	52
N50	16 06	16 09	16 12	16 16	16 20	16 24	16 28	16 33	16 38	16 43	16 48	16 53	16 58	17 03	17 08	N50
45	16 27	16 29	16 32	16 35	16 39	16 42	16 46	16 50	16 54	16 58	17 02	17 07	17 11	17 15	17 19	45
40	16 43	16 46	16 48	16 51	16 54	16 57	17 00	17 04	17 07	17 11	17 14	17 18	17 21	17 25	17 29	40
35	16 57	16 59	17 02	17 04	17 07	17 10	17 13	17 16	17 19	17 22	17 25	17 28	17 31	17 34	17 37	35
30	17 10	17 12	17 14	17 16	17 18	17 21	17 24	17 26	17 29	17 31	17 34	17 36	17 39	17 41	17 44	30
N20	17 30	17 32	17 34	17 36	17 38	17 40	17 42	17 44	17 46	17 48	17 50	17 51	17 53	17 55	17 57	N20
N10	17 49	17 50	17 52	17 53	17 55	17 57	17 58	18 00	18 01	18 02	18 03	18 05	18 06	18 07	18 07	N10
0	18 06	18 07	18 09	18 10	18 11	18 12	18 13	18 14	18 15	18 16	18 17	18 17	18 17	18 18	18 18	0
S10	18 23	18 25	18 26	18 27	18 28	18 28	18 29	18 29	18 30	18 30	18 30	18 30	18 29	18 29	18 28	S10
20	18 42	18 43	18 44	18 45	18 45	18 46	18 46	18 46	18 45	18 45	18 44	18 43	18 42	18 41	18 40	20
S30	19 04	19 05	19 05	19 06	19 06	19 05	19 05	19 04	19 03	19 02	19 01	18 59	18 57	18 55	18 53	S30
35	19 17	19 17	19 18	19 18	19 18	19 17	19 16	19 15	19 14	19 12	19 10	19 08	19 06	19 03	19 00	35
40	19 32	19 32	19 32	19 32	19 31	19 31	19 29	19 28	19 26	19 24	19 21	19 19	19 16	19 12	19 09	40
45	19 49	19 50	19 49	19 48	19 48	19 47	19 45	19 43	19 40	19 37	19 34	19 31	19 27	19 23	19 19	45
50	20 12	20 12	20 11	20 10	20 08	20 06	20 04	20 01	19 58	19 54	19 50	19 46	19 41	19 37	19 32	50
S52	20 22	20 22	20 21	20 20	20 18	20 16	20 13	20 10	20 06	20 02	19 58	19 53	19 48	19 43	19 37	S52
54	20 34	20 34	20 33	20 31	20 29	20 26	20 23	20 20	20 15	20 11	20 06	20 01	19 56	19 50	19 44	54
56	20 48	20 48	20 46	20 44	20 42	20 39	20 35	20 31	20 26	20 21	20 16	20 10	20 04	19 57	19 51	56
58	21 05	21 04	21 02	21 00	20 56	20 53	20 48	20 43	20 38	20 32	20 26	20 20	20 13	20 06	19 59	58
S60	21 25	21 23	21 21	21 18	21 14	21 09	21 04	20 59	20 52	20 46	20 39	20 31	20 24	20 16	20 08	S60

MORNING CIVIL TWILIGHT

Lat.	Dec. 30	Jan 2	Jan 5	Jan 8	Jan 11	Jan 14	Jan 17	Jan 20	Jan 23	Jan 26	Jan 29	Feb 1	Feb 4	Feb 7	Feb 10	Lat.
	h m	h m	h m	h m	h m	h m	h m	h m	h m	h m	h m	h m	h m	h m	h m	
N72	10 50	10 42	10 33	10 22	10 11	09 59	09 47	09 35	09 22	09 09	08 56	08 43	08 29	08 16	08 02	N72
70	09 53	09 49	09 44	09 38	09 31	09 23	09 14	09 05	08 55	08 45	08 34	08 23	08 12	08 01	07 49	70
68	09 19	09 17	09 13	09 09	09 04	08 57	08 51	08 43	08 35	08 27	08 18	08 08	07 59	07 48	07 38	68
66	08 54	08 53	08 50	08 47	08 43	08 38	08 32	08 26	08 19	08 12	08 04	07 56	07 47	07 38	07 29	66
64	08 35	08 34	08 32	08 29	08 26	08 22	08 17	08 12	08 06	07 59	07 53	07 45	07 38	07 30	07 21	64
62	08 19	08 19	08 17	08 15	08 12	08 08	08 04	08 00	07 55	07 49	07 43	07 36	07 29	07 22	07 15	62
N60	08 06	08 05	08 04	08 02	08 00	07 57	07 54	07 50	07 45	07 40	07 34	07 29	07 22	07 16	07 09	N60
58	07 54	07 54	07 53	07 52	07 50	07 47	07 44	07 40	07 36	07 32	07 27	07 22	07 16	07 10	07 04	58
56	07 44	07 44	07 44	07 42	07 41	07 38	07 36	07 33	07 29	07 25	07 20	07 15	07 10	07 05	06 59	56
54	07 36	07 36	07 35	07 34	07 33	07 31	07 28	07 25	07 22	07 18	07 14	07 10	07 05	07 00	06 55	54
52	07 27	07 28	07 27	07 26	07 25	07 23	07 21	07 19	07 16	07 12	07 09	07 05	07 00	06 56	06 51	52
N50	07 20	07 20	07 20	07 19	07 18	07 17	07 15	07 13	07 10	07 07	07 04	07 00	06 56	06 52	06 47	N50
45	07 04	07 05	07 05	07 04	07 04	07 03	07 01	07 00	06 58	06 55	06 53	06 50	06 46	06 43	06 39	45
40	06 51	06 52	06 52	06 52	06 52	06 51	06 50	06 49	06 47	06 45	06 43	06 41	06 38	06 35	06 32	40
35	06 39	06 40	06 40	06 41	06 41	06 40	06 40	06 39	06 38	06 37	06 35	06 33	06 31	06 29	06 26	35
30	06 29	06 30	06 30	06 31	06 31	06 31	06 31	06 30	06 29	06 29	06 27	06 26	06 24	06 22	06 20	30
N20	06 10	06 11	06 12	06 13	06 14	06 14	06 14	06 14	06 14	06 14	06 13	06 13	06 12	06 11	06 10	N20
N10	05 53	05 54	05 55	05 56	05 57	05 58	05 59	06 00	06 00	06 01	06 01	06 01	06 00	06 00	06 00	N10
0	05 36	05 38	05 39	05 40	05 42	05 43	05 44	05 45	05 46	05 47	05 48	05 48	05 49	05 49	05 49	0
S10	05 18	05 20	05 21	05 23	05 25	05 26	05 28	05 30	05 31	05 33	05 34	05 35	05 36	05 37	05 38	S10
20	04 58	05 00	05 02	05 04	05 06	05 08	05 10	05 12	05 14	05 16	05 18	05 20	05 22	05 24	05 26	20
S30	04 33	04 35	04 37	04 40	04 42	04 45	04 48	04 51	04 53	04 56	04 59	05 01	05 05	05 07	05 10	S30
35	04 18	04 20	04 23	04 25	04 28	04 31	04 34	04 38	04 41	04 44	04 48	04 51	04 54	04 58	05 01	35
40	04 00	04 02	04 05	04 08	04 11	04 15	04 18	04 22	04 26	04 30	04 34	04 38	04 42	04 46	04 50	40
45	03 38	03 40	03 43	03 47	03 51	03 55	03 59	04 03	04 08	04 12	04 17	04 22	04 27	04 31	04 36	45
50	03 08	03 11	03 15	03 19	03 24	03 28	03 33	03 39	03 44	03 50	03 56	04 02	04 08	04 14	04 19	50
S52	02 53	02 57	03 01	03 05	03 10	03 15	03 21	03 27	03 33	03 39	03 46	03 52	03 58	04 05	04 11	S52
54	02 36	02 39	02 44	02 49	02 54	03 00	03 06	03 13	03 20	03 27	03 34	03 41	03 48	03 55	04 02	54
56	02 14	02 18	02 23	02 29	02 35	02 42	02 49	02 57	03 04	03 12	03 20	03 28	03 36	03 44	03 52	56
58	01 45	01 50	01 56	02 03	02 11	02 19	02 28	02 37	02 46	02 55	03 04	03 13	03 22	03 31	03 40	58
S60	00 58	01 06	01 15	01 26	01 37	01 48	01 59	02 11	02 22	02 33	02 44	02 55	03 05	03 16	03 26	S60

EVENING CIVIL TWILIGHT

A63

Lat.	Dec. 30	2	5	8	11	14	17	20	23	26	29	1	4	7	10	Lat.
				January								February				
	h m	h m	h m	h m	h m	h m	h m	h m	h m	h m	h m	h m	h m	h m	h m	°
N72	13 15	13 26	13 38	13 51	14 05	14 19	14 33	14 48	15 03	15 17	15 31	15 46	16 00	16 14	16 28	N72
70	14 12	14 18	14 26	14 35	14 45	14 55	15 06	15 17	15 29	15 41	15 53	16 05	16 17	16 29	16 41	70
68	14 46	14 51	14 57	15 04	15 12	15 20	15 30	15 39	15 49	15 59	16 09	16 20	16 30	16 41	16 52	68
66	15 10	15 15	15 20	15 26	15 33	15 40	15 48	15 56	16 05	16 14	16 23	16 32	16 42	16 51	17 01	66
64	15 29	15 33	15 38	15 44	15 50	15 56	16 03	16 10	16 18	16 26	16 34	16 43	16 51	17 00	17 08	64
62	15 45	15 49	15 53	15 58	16 04	16 09	16 16	16 22	16 29	16 37	16 44	16 52	16 59	17 07	17 15	62
N60	15 58	16 02	16 06	16 11	16 15	16 21	16 27	16 33	16 39	16 46	16 52	16 59	17 06	17 13	17 21	N60
58	16 10	16 13	16 17	16 21	16 26	16 31	16 36	16 42	16 48	16 54	17 00	17 06	17 13	17 19	17 26	58
56	16 20	16 23	16 27	16 31	16 35	16 40	16 45	16 50	16 55	17 01	17 06	17 12	17 18	17 24	17 30	56
54	16 29	16 32	16 35	16 39	16 43	16 47	16 52	16 57	17 02	17 07	17 12	17 18	17 23	17 29	17 35	54
52	16 37	16 40	16 43	16 47	16 50	16 54	16 59	17 03	17 08	17 13	17 18	17 23	17 28	17 33	17 39	52
N50	16 44	16 47	16 50	16 53	16 57	17 01	17 05	17 09	17 14	17 18	17 23	17 28	17 32	17 37	17 42	N50
45	17 00	17 03	17 06	17 08	17 12	17 15	17 19	17 22	17 26	17 30	17 34	17 38	17 42	17 46	17 50	45
40	17 14	17 16	17 19	17 21	17 24	17 27	17 30	17 33	17 36	17 40	17 43	17 47	17 50	17 54	17 57	40
35	17 25	17 27	17 30	17 32	17 35	17 37	17 40	17 43	17 46	17 49	17 52	17 54	17 57	18 00	18 03	35
30	17 36	17 38	17 40	17 42	17 44	17 47	17 49	17 52	17 54	17 57	17 59	18 02	18 04	18 06	18 09	30
N20	17 54	17 56	17 58	18 00	18 02	18 04	18 06	18 08	18 09	18 11	18 13	18 14	18 16	18 18	18 19	N20
N10	18 12	18 13	18 15	18 16	18 18	18 19	18 21	18 22	18 23	18 24	18 26	18 27	18 28	18 28	18 29	N10
0	18 28	18 30	18 31	18 33	18 34	18 35	18 36	18 37	18 37	18 38	18 38	18 39	18 39	18 39	18 39	0
S10	18 46	18 48	18 49	18 50	18 50	18 51	18 52	18 52	18 52	18 52	18 52	18 52	18 51	18 51	18 50	S10
20	19 07	19 08	19 08	19 09	19 09	19 10	19 09	19 09	19 09	19 08	19 08	19 06	19 05	19 04	19 02	20
S30	19 31	19 32	19 32	19 33	19 33	19 32	19 32	19 31	19 29	19 28	19 26	19 25	19 22	19 20	19 18	S30
35	19 46	19 47	19 47	19 47	19 47	19 46	19 45	19 44	19 42	19 40	19 38	19 36	19 33	19 30	19 27	35
40	20 04	20 05	20 05	20 04	20 03	20 02	20 01	19 59	19 57	19 54	19 52	19 49	19 45	19 42	19 38	40
45	20 27	20 27	20 26	20 25	20 24	20 22	20 20	20 18	20 15	20 12	20 08	20 04	20 00	19 56	19 51	45
50	20 56	20 56	20 55	20 53	20 51	20 48	20 45	20 42	20 38	20 33	20 29	20 24	20 19	20 13	20 08	50
S52	21 11	21 10	21 09	21 07	21 04	21 01	20 57	20 53	20 49	20 44	20 39	20 34	20 28	20 22	20 16	S52
54	21 28	21 27	21 25	21 23	21 20	21 16	21 12	21 07	21 02	20 57	20 51	20 44	20 38	20 31	20 25	54
56	21 50	21 48	21 46	21 43	21 39	21 34	21 29	21 23	21 17	21 11	21 04	20 57	20 50	20 42	20 35	56
58	22 19	22 16	22 12	22 08	22 02	21 56	21 50	21 43	21 35	21 28	21 20	21 12	21 03	20 55	20 46	58
S60	23 05	22 59	22 52	22 44	22 36	22 26	22 17	22 08	21 59	21 49	21 39	21 29	21 20	21 10	21 00	S60

APPENDICES

Appendix 1: Construction of a Mercator

In the construction of a Mercator, two separate calculations are involved; calculation of spacing between the meridians and the calculation of spacing of the individual parallels.

It will be remembered that the scale along the meridians varies from point to point to give the chart orthomorphism. For the calculation of such infinitely small variations the calculus must be used. Meridional parts (mer parts or MPs) are handy sets of figures, giving the summation of each tiniest part of the meridian expanded by the secant of latitude. A mer part is defined as 'the number of times the chart length of 1' of longitude is contained in the chart length between the equator and the latitude in question'. Tabulated at 1' of ch lat, these tables may be used to construct a Mercator for training purposes.

Project

Construct a Mercator's chart covering 52°N to 56°N and 02°E to 02°W. Scale 1:1 000 000 at 54°N.

Construction

Step 1. First it is necessary to calculate the scale at the equator.

$$\text{Scale at the eq.} = \frac{1}{1\,000\,000 \times \sec 54^\circ}$$

$$= \frac{1}{1\,000\,000 \times 1.7013}$$

$$= \frac{1}{1\,701\,300}$$

Step 2. The next step is to find the chart distance representing a ch long of 1' (that is, 1 nm) at the equator. This will give us the chart distance between meridians 1° apart and this figure will also be used to calculate the latitude spacing. The mer parts of any latitude are the number of times the chart length of 1' of longitude is contained in the chart length between the equator and the latitude concerned.

$$CL = \text{scale} \times \text{earth distance}$$

$$= \frac{1}{1\,701\,300} \times 72\,960$$

$$= 0.043''$$

Step 3. This gives us the spacing of the meridians, 1° apart: 60 x 0.043 = 2.58".
Step 4. Now, to calculate the latitude spacing along a meridian, extract the information for the mer parts tables as necessary and complete the following table.

Lat.	Mer parts from the tables*	Difference in mer parts	x	CL of 1' of ch long	=	Chart distance
52°N	3646.74	—		—		—
53°N	3745.05	98.31		0.043		4.23″
54°N	3845.69	100.64		0.043		4.33″
55°N	3948.78	103.09		0.043		4.43″
56°N	4054.48	105.70		0.043		4.54″

* For the Terrestrial Spheroid, compression 1/293.465

We now have all the information we need to construct the graticule.

(a) Construct the base latitude, 52°N, as a horizontal straight line.

(b) Insert meridians at right angles to the base line, separation distance 2.58″ (as calculated in 3 above).

(c) Construct each latitude as a horizontal straight line parallel to the base using the chart length distance found in the last column of the table above.

Appendix 2: Lambert's Projection

The following project is for interest only – at no stage of your training does it become a part of the syllabus. But if you do attempt it you will find that you understand this projection a lot better. All you need is a sharp pencil, a set of four-figure tables and lots of scrap paper to do the sums. Check your answers at every stage with the answers given at the end.

Project
Construct a Lambert's orthomorphic projection covering 50N to 74N, 24W to 56W.

Further information
  (a) Scale of the chart is to be 1 inch to 300 nm.
  (b) The SPs are to be placed at 1/6th distance from the two limiting parallels.
  (c) Show meridians and parallels at 4° intervals.

Construction
Step 1. Locate the standard parallels. Where do you place the parallel of the origin?
Step 2. Calculate the radius of the reduced earth, R. Circumference $= 2\pi R$
Step 3. Calculate the value of the constant of the cone, n

$$ n = \frac{\log \sin \chi_1 - \log \sin \chi_2}{\log \tan \frac{\chi_1}{2} - \log \tan \frac{\chi_2}{2}} $$

where χ = co-lat; χ_1 = co-lat of SP_1; χ_2 = co-lat of SP_2.
Step 4. Calculate the position of the parallel of origin and comment on the result.
n = sine of the parallel of origin.
Step 5. Calculate the value of scale constant, K, for either standard parallel of your choice.

$$ K = \frac{R \sin \chi_1}{\tan^n \frac{\chi_1}{2} \times n} \quad \text{or} \quad \frac{R \sin \chi_2}{\tan^n \frac{\chi_2}{2} \times n} $$

Step 6. Calculate the radii in inches of the parallels, at 4° intervals, from 50°N upwards. Tabulate the calculated values in the table using

$$ R = K \tan^n \frac{\chi}{2} $$

(Notice that the scale constant K found for one of the two standard parallels is used in finding the radius of the other parallels.)

Step 7. Calculate difference between the radii of successive parallels and enter the findings in the table.

Step 8. Study the figures in the column under 'difference' and comment.

Step 9. Draw a central meridian as a vertical straight line in the centre of the paper. From a convenient point on the extension of this meridian, draw in the parallels as arcs of circles at the calculated distances. Number the parallels, indicate the two standards and the parallel of origin.

Step 10. Calculate the spacing of the other meridians from the central meridian on both standard parallels at the correct scale and mark off the distances on the two parallels.

$$\text{Spacing on } SP_1 = \frac{2\pi R \cos SP_1}{360°} \times \text{ch long}$$

$$\text{Spacing on } SP_2 = \frac{2\pi R \cos SP_2}{360°} \times \text{ch long}$$

Step 11. Draw in the meridians by joining the equivalent points marked off on the two SPs and number them.

Step 12. Check the scale factor *SF* for each parallel shown on the graticule.

$$SF = \frac{\sin \chi_1 \times \tan^n \frac{\chi}{2}}{\sin \chi \times \tan^n \frac{\chi_1}{2}}$$

Enter the results in the table.

Step 13. Study the *SF* for each parallel and comment on the scale aspect of the property of the projection.

Parallel	Radius	Difference	SF
50°N		—	
54°N			
58°N			
62°N			
66°N			
70°N			
74°N			

Answers
Step 1: 54°N and 70°N
Step 2: 11.46 inches
Step 3: $n = 0.8857$
Step 4: Parallel of origin at 62°21′, north of the midway between the two SPs
Step 5: $K = 20.59$
Steps 6, 7 and 12:

Parallel	Radius	Difference	*SF*
50°N	8.412"	–	1.011
54°N	7.607"	0.805	1.000
58°N	6.803"	0.804	0.993
62°N	6.017"	0.786	0.990
66° N	5.224"	0.793	0.993
70° N	4.427"	0.797	1.000
74° N	3.621"	0.806	1.015

Step 10: SP_1 = 0.470
  SP_2 = 0.274

Step 13: The scale is correct at the two standard parallels: contracts inward, expands outward. The maximum contraction occurs at the parallel of origin. The expansion is slightly higher at the northern limit of the chart, that is 74°N, than at the southern limit, 50°N.

Appendix 3: Glossary of Abbreviations

a/c	aircraft
amsl	above mean sea level
agl	above ground level
A/H	alter heading
alt or Alt	altitude
ASI	airspeed indicator
ASR	Altimeter Setting Region
ATD	actual time of departure
Brg	Bearing
°C	degrees Celsius, hitherto called Centigrade
°(C)	degrees Compass
CA	conversion angle
CAA	Civil Aviation Authority
CAVOK	weather fine and clear
CDU	control display unit
CH	celestial horizon
ch lat	change of latitude
ch long	change of longitude
CL	chart length
cm	centimetre(s)
CM	Central Meridian
C of G	centre of gravity
COAT	corrected outside air temperature
C/S or c/s	call sign
CP	critical point
CRV	Centre reading voltmeter
Dev	deviation
DGI	directional gyro indicator
dist	distance
DME	distance measuring equipment
DR	dead reckoning
ED	Earth distance
EMF	electro-motive force
ETA	estimated time of arrival
FL	flight level
ft	feet
ft/min	feet per minute
°(G)	degrees Grid

GC	Great circle
GD	Greenwich date
GMT	Greenwich mean time
GN	Grid North
Griv	grivation
G/S	ground speed
Hdg	Heading
HHI	horizontal hard iron
h, m, s	hours, minutes, seconds
h	hour(s)
ht	height
IAS	indicated airspeed
IFR	instrument flight rules
in	inch
INS	Inertial Navigation System
ISA	International Standard atmosphere
kg	kilogram(s)
kg/h	kilograms per hour
km/h	kilometres per hour
kt	knot(s)
Lat	Latitude
LD	Local Date; also landing distance
LMT	Local Mean Time
Long	Longitude
LV	light and variable
M	Mach
$^\circ$(M)	degrees Magnetic
Mb/mb	millibar(s)
Mcrit	critical Mach number
MI	Master indicator
min	minute(s)
M_{ind}	indicated Mach number
mm	millimetre(s)
MN	Mach number, Magnetic north
mph	statute miles per hour
msl	mean sea level
M/R	Moonrise
M/S	Moonset
NH	Northern hemisphere
NM/nm	nautical mile(s)
NP	North Pole
PE	pressure error
P/L	position line
posn	position
Press Alt	pressure altitude
QDM	Hdg(M) to station with zero wind
QDR	Brg (M) from station

QFE, QFF, QNH	defined in text
QNE	altimeter reading with 1013.2 mb set
QTE	Brg (T) from station
RAS	rectified airspeed
Rel	relative
RL	rhumb line
rpm	revolutions per minute
R/W	Runway
SH	Southern Hemisphere
S/H	set heading
sm	statute mile(s)
SP	South Pole
S/R	Sunrise
S/S	Sunset
ST	Standard Time
°(T)	degrees True
TAS	True airspeed
Temp	temperature
TMG	Track made good
TN	True North
T/O	take-off
TOC	top of climb
Tr	Track
Var	variation
VSC	Variation setting corrector
VSI	Vertical Soft Iron, also Vertical Speed Indicator
WPT	Waypoint
W/S	wind speed
W/V	wind velocity

Index